AF769016

Educating for a Climate Changed Future

Fernando Reimers · Margaret Wang-Aghania

Educating for a Climate Changed Future

From First Order Impacts to System Level Transformation

Springer

Fernando Reimers
Harvard Graduate School of Education
Cambridge, MA, USA

Margaret Wang-Aghania
SubjectToClimate
Watertown, MA, USA

ISBN 978-3-032-23255-7 ISBN 978-3-032-23256-4 (eBook)
https://doi.org/10.1007/978-3-032-23256-4

© The Editor(s) (if applicable) and The Author(s) 2026. This book is an open access publication.

Open Access This book is licensed under the terms of the Creative Commons Attribution-NonCommercial-NoDerivatives 4.0 International License (http://creativecommons.org/licenses/by-nc-nd/4.0/), which permits any noncommercial use, sharing, distribution and reproduction in any medium or format, as long as you give appropriate credit to the original author(s) and the source, provide a link to the Creative Commons license and indicate if you modified the licensed material. You do not have permission under this license to share adapted material derived from this book or parts of it.
The images or other third party material in this book are included in the book's Creative Commons license, unless indicated otherwise in a credit line to the material. If material is not included in the book's Creative Commons license and your intended use is not permitted by statutory regulation or exceeds the permitted use, you will need to obtain permission directly from the copyright holder.
This work is subject to copyright. All commercial rights are reserved by the author(s), whether the whole or part of the material is concerned, specifically the rights of reprinting, reuse of illustrations, recitation, broadcasting, reproduction on microfilms or in any other physical way, and transmission or information storage and retrieval, electronic adaptation, computer software, or by similar or dissimilar methodology now known or hereafter developed. Regarding these commercial rights a non-exclusive license has been granted to the publisher.
The use of general descriptive names, registered names, trademarks, service marks, etc. in this publication does not imply, even in the absence of a specific statement, that such names are exempt from the relevant protective laws and regulations and therefore free for general use.
The publisher, the authors and the editors are safe to assume that the advice and information in this book are believed to be true and accurate at the date of publication. Neither the publisher nor the authors or the editors give a warranty, expressed or implied, with respect to the material contained herein or for any errors or omissions that may have been made. The publisher remains neutral with regard to jurisdictional claims in published maps and institutional affiliations.

This Springer imprint is published by the registered company Springer Nature Switzerland AG
The registered company address is: Gewerbestrasse 11, 6330 Cham, Switzerland

If disposing of this product, please recycle the paper.

Contents

Chapter 1
Introduction: Education and Climate

This book examines the intricate and evolving relationship between education and climate change from a global and systemic perspective. We treat "climate" and "education" as dynamic, interacting systems whose interactions are disrupting education structures, cultures and purposes and preparing students to adapt to mitigate climate change.

Climate change is now widely recognized as "widespread, rapid, and intensifying," unequivocally human-induced, and already producing irreversible changes in many natural and social systems (IPCC, 2014; 2022a, 2022b). In 2025, scientists from the Massachusetts Institute of Technology reported that the planet was on track to exceed key climate thresholds, which would result in predictable deaths for over half a million people a year, increasing extreme weather, erratic precipitation resulting in flooding and droughts (Paltsev et al., 2025). At the same time, the world has recently experienced a global pandemic that disrupted learning for hundreds of millions of students (Reimers, 2024). These crises are not merely parallel; they expose shared patterns of unsustainable development and unequal vulnerability, and they open a window to reimagine the purposes and practices of education to support more sustainable futures. Recent synthetic reviews underscore this bi-directional view, documenting how climate change already affects schooling and how education may, in turn, shape environmental outcomes, while highlighting how limited the causal evidence still is (Ambasz et al., 2025; CLEAN, 2025; Kraft et al., 2025). At the same time, an emerging body of work on climate change education has begun to define the field conceptually and practically, emphasizing the need to connect scientific understanding with action, identity, and justice (Chang, 2014, 2022; Singh, 2023; Xu & Iyengar, 2023).

While we see education as an avenue to help learners gain knowledge and skills that can support mitigation and adaptation to climate change, we do not assume that education can "solve" climate change by itself. The drivers of climate disruption lie in economic structures, political choices, and technological systems that extend far beyond schools. Our purpose is more modest and more realistic: to document how

© The Author(s) 2026

F. Reimers and M. Wang-Aghania, *Educating for a Climate Changed Future*,
https://doi.org/10.1007/978-3-032-23256-4_1

education and climate change currently shape each other; to analyze the narratives, policies and practices that link them; to assess what is known about the effectiveness of those efforts; and to identify where education systems could be doing more, and doing things differently, to support mitigation, adaptation, and just transitions. This agenda resonates with emerging research that organizes the education–climate nexus around multiple, interrelated dimensions, including schools' environmental footprints, their resilience to climate risks, supports for affected students and staff, teaching and learning about climate and sustainability, and schools' roles as community hubs for climate awareness and action (Kraft et al., 2025), as well as with scholarship that conceptualizes climate change education (CCE) as a distinct but closely related field within environmental and sustainability education (Chang, 2014, 2022; Xu & Iyengar, 2023).

Around the world, the work of teachers and students is already being transformed by rising temperatures, changing precipitation patterns, and more frequent extreme weather events. These processes, related to climate change, affect teaching and learning directly and indirectly. Directly, climate hazards close schools, damage buildings, disrupt instructional time, and impair students' health, safety and cognitive functioning. Indirectly, climate shocks and slow-onset changes alter livelihoods, migration patterns, public finances and social stability, amplifying inequalities that shape educational opportunity. Syntheses of recent evidence for both low- and high-income countries document how these first-order impacts of climate change on schooling have become systemic and recurrent rather than exceptional (Kraft et al., 2025; Sabarwal et al., 2024). In January of 2025 UNICEF estimated that climate related events disrupted education for 250 million children in 84 countries the previous year, 74% of them in low and middle income countries (UNICEF, 2025).

At the same time, climate change is prompting new educational responses. Governments, schools, teachers, supranational organizations and grassroots movements are revising curricula, experimenting with pedagogies, building climate-resilient infrastructure, rethinking school operations, and advocating for "green skills" and climate literacy. Some of these efforts frame climate education mainly as scientific literacy; others emphasize sustainable development, green economies, or justice-centered, decolonial approaches. A growing, but still limited, body of research examines whether and how these educational interventions influence environmental knowledge, attitudes and behaviors, with recent reviews emphasizing both the promise and the difficulty of generating durable behavior change through education (Ambasz et al., 2025; Ardoin et al., 2020). Within this emerging field, Chang's *Climate Change Education: Knowing, Doing and Being* (2014, 2022) has been influential in arguing that CCE must attend simultaneously to understanding ("knowing"), action ("doing") and identity ("being"), while Xu and Iyengar's *Climate Change Education: An Earth Institute Sustainability Primer* (2023) and Singh's *Teaching Climate Change: Science, Stories, Justice* (2023) further highlight the importance of systems thinking, stories, justice and climate emotions in climate pedagogy.

This book seeks to make sense of these developments through a comparative and multidisciplinary lens. It traces the multiple channels through which climate change and education interact, and it asks three sets of questions:

- Conceptualization. How has the relationship between education and climate change been understood? In what ways are climate and education interdependent systems, each shaping the other? How do different narratives, climate literacy, education for climate action, green skills, education for sustainable development, critical and decolonial climate education, frame education's role?
- Practice. How do these aims translate into national systems, informal networks, schools, classrooms, and ultimately, student outcomes? In other words, what approaches exist in practice and what evidence do we have about their effects on learning, behavior, policy and climate outcomes?
- Gaps and opportunities. Where are the mismatches between ambitions and reality? Which learners, regions and perspectives are most neglected? What opportunities exist to align education systems more closely with the demands of a rapidly warming world?

To answer these questions, we synthesize empirical knowledge from multiple sources: global reviews of policy and curriculum, comparative research on education systems, assessment data, case studies of schools and communities, and emerging evidence on climate impacts on learning. Our approach is informed by a systems view: we treat education and climate as complex, adaptive systems in interaction, and we attend to how their relationship is mediated by governance, culture, professional practice, political economy and infrastructure. We also draw on recent conceptual frameworks that specify cognitive, affective and situational pathways through which schooling can influence environmental outcomes, and that call attention to the predominance of correlational over causal studies at this intersection (Ambasz et al., 2025), as well as on work in climate change education that argues for integrating "knowing, doing and being" (Chang, 2014, 2022), transdisciplinary and justice-centered approaches (Singh, 2023), and systems thinking across formal and informal learning contexts (Xu & Iyengar, 2023).

Building on established frameworks for policy coherence in education, as well as Reimers' (2020a, 2020b) multi-perspective model for educational change, we conceptualize education as not only a domain where climate impacts and adaptive responses manifest, but also as an integral component of larger ecological, social, and political systems that collectively influence resilience and enable collective action. The policy coherence framework, as articulated by Darling-Hammond (2010), Elmore (1996), and further developed by the OECD (2010, 2013), emphasizes the importance of alignment and of mutually reinforcing nature of policies across sectors, levels of governance, and domains of practice, ensuring that educational goals, strategies, and resources work synergistically to promote desired outcomes and mitigate contradictory effects. This approach is particularly relevant to complex challenges such as climate change, where education policies must integrate multiple goals, such as civic preparation or workforce development and where policies in education must be systematically coordinated with those in environmental governance, health, and social welfare to build adaptive capacity and sustainable futures.

We draw explicitly on global frameworks developed by UNESCO and its partners, including the Belgrade Charter and the Tbilisi Intergovernmental Conference on Environmental Education (UNESCO, 1975, 1977), Agenda 21 (United Nations Conference on Environment and Development [UNCED], 1992), and subsequent guidance such as UNESCO's ESD Lens review tool (UNESCO, 2010a, 2010b), documents from the UN Decade of Education for Sustainable Development 2005–2014, The DESD at a glance (UNESCO, 2005), the UNESCO roadmap for implementing the Global Action Program on ESD (UNESCO, 2014a, 2014b), Education for Sustainable Development Goals: Learning Objectives (UNESCO, 2017a, 2017b), and Issues and Trends in Education for Sustainable Development (Leicht et al., 2018). We also build on the ESD for 2030 framework and roadmap (UNESCO, 2019, 2020), the global review Learn for our planet (UNESCO, 2021a, 2021b, 2021c, 2021d), the Berlin Declaration on ESD (UNESCO, 2022), and recent curriculum work on climate action (Iyengar & Kwauk, 2021a, 2021b). Together, these frameworks and reviews conceptualize "green education" through interconnected dimensions that include inclusive and participatory governance; transformative teaching, learning and assessment; communitarian lifelong learning settings; and sustainable, locally pertinent infrastructure (Iyengar & Kwauk, 2021a, 2021b; Leicht et al., 2018; UNCED, 1992; UNESCO, 1975, 1978, 2005, 2010a, 2010b, 2014a, 2014b, 2017a, 2017b, 2019, 2020, 2021a, 2021b, 2021c, 2021d, 2022). Parallel to this institutional architecture, syntheses such as Chang (2014, 2022), Singh (2023), Xu and Iyengar (2023), Lê (2025) and Worth (2021) help delineate CCE as a scholarly and practical domain, clarifying how climate change education relates to environmental education and ESD, and what distinctive emphases, systems thinking, justice, stories, teacher agency, and political economy, it adds.

We see climate change as the defining challenge of our century, and the institutions of education as crucial in shaping how humanity responds. While climate disrupts schools, education itself is also a powerful lever for adaptation, resilience and mitigation. Our central argument is that education and climate are mutually constitutive systems: each is both "cause" and "effect" of the other, and each constrains and enables the possibilities of the other. This view is consistent with recent interdisciplinary syntheses that characterize the relationship between schooling and climate as multi-dimensional, spanning infrastructure, operations, pedagogy, student and staff well-being, and community roles, and emphasize that schools are at once vulnerable to climate risks and uniquely positioned to contribute to solutions (Kraft et al., 2025).

To develop this argument, we introduce three foundational ideas in this chapter. First, we adopt a systems perspective, explaining how education and climate can be understood as complex systems in interaction. Second, we emphasize timescales, showing how both climate change and educational change unfold over decades and across the life course. Third, we introduce a distinction between first-order effects (how climate change disrupts education) and second-order effects (how education systems respond to, and act upon, climate change), including the emerging landscape of "green education" and global climate-education governance. These lenses complement recent conceptual work that specifies the cognitive, affective and situational mechanisms through which education can influence environmental attitudes and behaviors (Ambasz et al., 2025), and they connect with CCE frameworks that insist on linking knowledge to action and identity (Chang, 2014, 2022; Singh, 2023; Xu & Iyengar, 2023).

The rest of the book builds on these foundations, exploring how climate change and education interact across systems, institutions, and communities. Later chapters map the terrain of perspectives on education and climate change, review evidence on what students know and how climate is taught, analyze system-level levers such as curriculum, assessment and teacher development, and examine the role of informal networks and communities. Throughout, we adopt both a systems approach and an equity lens, recognizing that the impact of climate change on educational opportunity is deeply unequal across and within countries, as are the opportunities to gain knowledge and skills to respond to climate change.

1.1 A System's View of Climate, Education and Climate Change

Educational institutions are embedded in climate systems, and therefore are impacted by them, while also being capable of impacting the climate system. In this sense, education and climate are mutually constitutive: each is both "cause" and "effect" of the other. These interactions unfold over long periods of time. The climatic changes that currently disrupt schools and reshape the purposes of education have roots in human activities decades or centuries ago; conversely, the knowledge, skills and values that schools transmit will influence emissions, adaptation choices and political decisions mostly over the long term. Understanding these long-range, reciprocal influences requires a systems perspective.

A system can be understood as a set of components in relationship with one another through multiple interconnections. Those components and their interconnections shape a complex whole with properties and behaviors that cannot be explained by the components in isolation, but only by their relationships. The structure of a system comprises the organizational arrangements of components and the pattern of their interactions. Boundaries specify what is considered internal or external to the system. The environment in which a system is embedded influences the system and is influenced by it yet is not under its direct control. Systems exhibit characteristic patterns of behavior that emerge from these interactions, not from any single element alone. Feedback loops, both reinforcing and balancing, allow for self-regulation, adaptation and evolution in response to environmental changes.

Contemporary complexity science has deepened this basic intuition. In complex systems, relationships between cause and effect are often non-linear and sensitive to initial conditions: small inputs or perturbations can, under some circumstances, produce disproportionately large impacts because of feedback loops and networked interdependencies (Mitchell, 2009). New patterns of behavior can emerge from simple local interactions among components without any central controller, a phenomenon known as emergence (Holland, 1998). Complex systems also adapt and reorganize in response to external shocks or internal tensions (Waldrop, 1992). Their dynamics are shaped by reinforcing (positive) and balancing (negative) feedback loops that can stabilize or destabilize the system (Meadows, 2008). No single part has complete control as information and influence are distributed (Bar-Yam, 1997). Thinking in systems terms therefore requires us to "see the forest and the trees" at the same time: to attend both to individual elements and to the patterns of interaction that produce aggregate outcomes.

Examples of systems include biological organisms and their constituent subsystems. A plant is a living system comprising roots, stems, leaves and flowers whose coordinated interactions absorb water, sunlight and nutrients and carry out photosynthesis. The digestive system of a mammal transforms food into nutrients for the organism through the interconnections of mouth, esophagus, stomach, intestines, liver and pancreas. A colony of ants is a social system in which individual ants perform roles such as foraging, cleaning or defending the colony, and through their local interactions sustain the survival and reproduction of the colony. The solar system is defined by the sun, planets and other bodies, and the gravitational relationships that determine their motion.

Human-made artifacts can also be systems. A bicycle connects frame, wheels, chain, pedals and seat so that human effort can be converted into motion. Cars and airplanes link many more components into coherent structures that support transportation over much larger distances. Their performance depends not on any single part, but on how all parts are configured and interact.

Schools, like bicycles or airplanes, are human-made systems. They bring together buildings, students, teachers, instructional resources, routines and rules in ways that enable opportunities to learn. On a broader scale, a public education system, at the level of a nation or sub-national jurisdiction, encompasses the full organizational infrastructure through which a society intentionally passes on knowledge, skills and values to the next generation. This infrastructure includes institutions (schools, universities, ministries), policies and regulations, curricula, assessment and examination systems, funding flows, teacher education and professional development, buildings and equipment. It is helpful to think of an education system not as a static platform but as a dynamic, evolving entity, where overall direction and functioning are the result of interactions among its components rather than the simple sum of their individual effects. Just as the contribution of a bicycle wheel to movement can only be understood by reference to how it relates to the other components of the bicycle, the contribution of a teacher's instruction on climate change can only be understood in relation to curriculum, assessments, school leadership, available resources, governance structures and broader policy environments.

The interdependence of curriculum, teachers and assessment illustrates this point. Curriculum, embodying the goals, scope and sequence of desired learning, is not self-executing. It only becomes real through what teachers understand, value and are able to enact with students, using particular materials in specific classrooms. If curriculum reforms introduce climate change concepts, systems thinking or project-based pedagogies that are far beyond many teachers' current knowledge or practices, professional development and support will be required for those reforms to translate into classroom change (Chang, 2014, 2022; Lê, 2025). Likewise, assessment systems that continue to focus narrowly on factual recall will tend to steer teaching away from complex climate-related competencies, even if curricula have been revised. Coherent alignment across these elements, curriculum, teacher capacity, assessment, emerges when they are designed with attention to one another; misalignment can undermine well-intended reforms. In this sense, what happens in a single classroom is always conditioned by the wider system: regulations, accountability pressures, teacher labor markets, funding, and also by transnational norms and networks.

Education systems are themselves nested both within larger national systems of government and within systems of global education governance and shaped by ideas that travel across national boundaries. The United Nations (UN), created after World War II, is a complex supranational system comprising multiple specialized agencies that carry out functions of global governance and international cooperation, including the advancement of peace and security. Among these, the United Nations Educational, Scientific and Cultural Organization (UNESCO) and the United Nations Children's Fund (UNICEF) support governments in realizing the universal right to education and the purposes set out in Article 26 of the Universal Declaration of Human Rights, which affirms that everyone has the right to free and compulsory elementary education, that education shall be directed to the full development of the human personality and to the strengthening of respect for human rights and fundamental freedoms, and that parents have a prior right to choose the kind of education that shall be given to their children.

UNESCO and UNICEF advance the right to education through several interrelated functions:

(a) Generating and disseminating ideas, anticipating and responding to emerging trends and needs in education, and supporting the development of education policies based on research and country priorities;
(b) Developing and promoting the adoption of education standards, shaping policies and practices that define what counts as high-quality and equitable education;
(c) Serving as clearinghouses, promoting the development, implementation and dissemination of successful educational policies and practices, setting norms and standards, and supporting their implementation;
(d) Building capacity, providing technical cooperation to develop the ability of member states to achieve their national education goals; and
(e) Catalyzing international cooperation, initiating and promoting dialogue and exchange among education leaders and stakeholders.

These UN organizations, together with other supranational actors such as the Organization for Economic Cooperation and Development, the World Bank, regional development banks and cooperation agencies, and bilateral aid agencies, form an architecture that enables global governance in education. In effect, they integrate education systems around the world into a "world system." Stanford Professor John Meyer and his colleagues have shown how a growing global consensus about human rights has influenced the worldwide expansion of mass education, as well as the content of school curricula, including the incorporation of human rights education, the emphasis on educating citizens with rights and responsibilities, and a greater focus on civics education (Meyer, 1977; Meyer et al., 1992, 2010). Later in this chapter, we will explain how global education governance has similarly contributed to the expansion and rapid evolution of efforts to align education curriculum with climate change.

In addition to intergovernmental organizations, there are also non-governmental educational organizations that seek to influence and support national systems. These include secular or religious networks of schools, organizations that shape curricula such as the International Baccalaureate, and global federations like Teach for All that engage recent university graduates in teaching disadvantaged students. There are also organizations that spotlight and recognize climate change education programs, thus contributing to the diffusion of innovation in this domain, such as the organizations Hundred.org or T4education.

Just as education institutions can be understood as systems, so can climate. Climate refers to the long-term patterns in weather variables, such as temperature, humidity, atmospheric pressure, wind and precipitation, typically characterized over periods of about 30 years or more (IPCC, 2001; WMO, 2022). These patterns are shaped by factors including latitude, altitude, topography, distance from oceans, ocean currents, land cover and land use, as well as external forces such as greenhouse gas concentrations, changes in solar radiation and volcanic eruptions (IPCC, 2001, 2018). In turn, climate regulates key conditions for life, including temperature ranges and water availability, and thereby influences the prospects for human societies and other living organisms (IPCC, 2018; WMO, 2022).

The climate system emerges from interactions among several major components: the atmosphere, the hydrosphere (including the oceans and other water bodies), the cryosphere, the land surface and the biosphere (IPCC, 2001). Natural processes and human activities, especially increasing greenhouse gas emissions and land-use change, can modify these interactions and alter climate over time (IPCC, 2018). As a complex system, climate exhibits non-linear responses, thresholds and tipping points, along with feedback loops such as the ice–albedo feedback, in which the loss of reflective ice surfaces accelerates additional warming (IPCC, 2018; United Nations Climate Security Mechanism, n.d.). These dynamics mean that relatively small perturbations can sometimes give rise to disproportionately large and difficult-to-reverse changes in the climate system.

Climate systems form part of the external environment in which education systems operate. Research shows that higher temperatures during the school year can reduce learning and test performance, with extreme heat having particularly harmful effects; school infrastructure such as air conditioning can mitigate some of these impacts (Park et al., 2020; 2021). Climate-related hazards, such as storms, floods and droughts, damage school infrastructure, disrupt attendance and displace learners, leading to lost instructional time and increased risk of dropout (Sabarwal et al., 2024; UNESCO, 2023). Climate-induced displacement also creates additional barriers to education, including overcrowded schools, administrative and legal obstacles, and heightened vulnerability among affected learners (UNESCO, 2023). Syntheses of evidence from the United States underline that climate-related school closures, infrastructure damage, rising insurance costs and unsafe heat and air-quality conditions are now routine features of the K–12 landscape, especially for already disadvantaged communities (Kraft et al., 2025). In response to these interdependencies, international initiatives increasingly call for integrating climate change into curricula so that students develop the knowledge, skills, values and dispositions needed to understand climate science, contribute to mitigation and adaptation, and engage civically with climate policy (UNESCO, 2021a, 2021b, 2021c, 2021d, 2024a, 2024b, 2024c).

Educational responses to climate may, in turn, shape climate outcomes, particularly over the long term. Education institutions shape climate in that school infrastructure, transportation and the goods and technologies consumed in schools have an environmental footprint encompassing energy use in buildings, emissions from commuting and school transport, resource- and carbon-intensive food provision, and waste and water use associated with daily operations. Growing use of digital technologies, including AI tools in classrooms and administration, further links schools to the rising energy and water demands of data centers, whose climate impact depends on underlying electricity mixes and cooling systems. Older, inefficient buildings, car-dependent access patterns, meat-heavy cafeterias and resource-intensive ICT use can significantly increase greenhouse gas emissions and other environmental pressures, while procurement of textbooks, digital devices, furniture and cleaning products embeds additional "upstream" impacts in education supply chains. At the same time, schools can serve as sites of mitigation and adaptation by investing in low-carbon and climate-resilient infrastructure, reducing water waste, shifting towards more plant-rich diets in cafeterias, improving building efficiency, promoting more sustainable transportation systems, and making careful, efficiency-oriented choices about digital and AI adoption—using campus operations (e.g., energy management, waste reduction, school gardens and mobility plans) as applied learning environments that reinforce pro-environmental behaviors. Education responses such as teaching about climate change, sustainability and alternative development pathways, may shift behavior related to consumption, energy use, transportation and civic engagement, which in turn influence emissions trajectories and adaptive capacities.

More generally, education institutions impact climate in that more educated people can better adapt and mitigate their impact on climate. For example, expanding girls' access to quality education tends to increase their agency over reproductive decisions and is associated with lower fertility rates; since population growth is one of the most impactful human contributors to climate change, it follows that girls education contributes to mitigate climate change. A recent review of empirical studies on education and environmental behavior concludes that schooling is consistently associated with greater environmental concern and, in some contexts, with more sustainable practices, though rigorous causal evidence remains scarce and effects on behavior appear more elusive than effects on attitudes (Ambasz et al., 2025). Education systems are therefore in interaction with climate systems, insofar as they shape how humans live and act, and insofar as educational infrastructure and supply chains (for example, school construction, energy use, food provision and transportation of students) have direct environmental and climatic impacts (Kraft et al., 2025).

For these reasons, understanding the relationship between education and climate requires a systems approach: a perspective that views this relationship as the outcome of interactions between two complex systems, education and climate, each composed of interconnected subsystems, and each forming part of the external environment of the other. The full contributions of any element, a classroom lesson, a national examination, a local weather pattern, an international climate agreement, can only be understood by reference to the systems of which it is a part. This requires understanding how different parts influence one another and collectively shape the overall nature of the relationship between education and climate. The goal of this book is to illuminate these relationships. However, before examining them in greater depth, we must first address the different timescales at which climate and education change.

1.2 Timescales of Climate and Educational Change

As already mentioned, climate refers to long-term patterns in weather; understanding it requires distinguishing long-term trends ("signal") from shorter-term variations ("noise") (Hulme, 2014). The long-term nature of climate change means that many of its proximate "causes" lie in emissions and other activities in the more distant past, and that present changes in human production and consumption patterns will exert most of their influence on the climate system only in the future (Hulme, 2014; Oppenheimer & Petsonk, 2005). In the near term, some changes in climate that we will observe are already effectively committed by past and current concentrations of greenhouse gases in the atmosphere, given the inertia of the climate system (Oppenheimer & Petsonk, 2005). Understanding climate therefore requires attending carefully to the different timescales on which climate processes and responses unfold (Risbey, 2008).

The effects of climate on individuals also accumulate across the life course. Exposure to climate-related hazards and environmental stressors at sensitive developmental periods, such as early childhood, can undermine cognitive, emotional and physical development, with consequences that compound over time and shape later vulnerability to subsequent climate events (National Research Council, 1998). This means that climate change can have both immediate and long-term, cumulative impacts on human development, making it particularly important for education systems and other social institutions to account for these temporal dimensions.

The changes that education could possibly induce in human behavior to impact climate, in turn, also play out over the long term. While some of the results of education efforts are immediate, students attending school, or learning what is intended in the curriculum, the most consequential results of those efforts play out over time, in the form of the consequences to individuals and to their societies of having educated their students in particular ways. For example, the diminished number of offspring that results from educating girls has an impact on climate over the long term, as do the changes in consumption and lifestyle that individuals can adopt as a result of greater knowledge and dispositions to act in ways that diminish their carbon footprint, such as shifting to a plant-based diet. In addition, some of the effects of education about climate also take place over the life course of students, where knowledge in early years supports subsequent learning. Understanding the impact of educational change is therefore also about understanding the timescales and developmental trajectories at which it operates.

The fact that both climate change and education are processes with long-term results, and that a life-course perspective is necessary to understand how they impact human development, creates an additional layer of complexity in understanding the relationship between these two systems. When education institutions are being affected by climate, they are being affected by events whose root causes are in the distant past. When the same institutions educate students to adapt to and mitigate climate, many of the results of those efforts will play out well into the future. Educating students so they can adapt to and mitigate the impact of climate is therefore largely about educating them for a changing future, a future they may not directly experience. This temporal mismatch also complicates efforts to evaluate climate-related education initiatives, since many of the hoped-for environmental benefits lie beyond the time horizons of most research designs, a difficulty that recent reviews of education and environmental outcomes explicitly acknowledge (Ambasz et al., 2025; Kraft et al., 2025).

1.3 The Relationship Between Education and Climate

There are first- and second-order effects of climate change on education. First-order effects are the influences of climate change on students, teachers or institutions that disrupt the process of education. Responses to these first-order effects aim at mitigating the disruption to learning caused by climate. Second-order effects include changes in the curriculum and other school processes, and in school operations more generally, with the aim of preparing students to understand climate change, and to equip them with the skills that will mitigate the disruptions that climate change will cause to their lives, including supporting the transition to a green economy and more sustainable forms of life. These second-order responses may also include changes in the management and operations of schools to reduce their impact on climate.

This distinction parallels, but is not identical to, recent typologies that organize the education–climate relationship around multiple dimensions: reducing schools' environmental impacts; making schools more resilient to environmental pressures; supporting students and staff affected by climate change; teaching about climate science, nature and green jobs; and expanding schools' role as community hubs for climate awareness and care for the planet (Kraft et al., 2025). Our first-order effects correspond roughly to the vulnerabilities and resiliency of school systems, while second-order effects encompass both instructional and institutional responses that position education as part of broader climate solutions. Within the second-order domain, Chang (2014, 2022) has framed climate change education in terms of "knowing, doing and being": developing robust climate understanding ("knowing"), enabling and supporting individual and collective action ("doing"), and cultivating identities, values and dispositions aligned with environmental responsibility and global citizenship ("being"). This triadic framing underscores that second-order effects concern not only what students learn about climate change, but also how they are enabled to act and who they are encouraged to become.

Examples of first-order effects include the impact on human populations, school infrastructure and operations of rising temperatures, changing precipitation patterns, and more frequent extreme weather events like hurricanes, wildfires, and droughts. In addition, they include responses such as adaptations to school functioning to mitigate such impact, for instance, increasing the resiliency of school infrastructure to floods, or developing hybrid modalities of instruction or flexible school schedules to maintain instructional time in the face of extreme weather events, or providing accommodations to enable climate migrants to continue their education.

Examples of second-order effects include educating girls, and boys, which reduces the age at which women become mothers, thus curbing overall fertility rates and consequently diminishing population growth which has a direct mitigating impact on climate change. In addition, schools can help students understand the process of climate change, how it is impacting the localities where they live, and what kind of actions might be adaptive or mitigate impact. Schools can also help students gain the skills that would help transition to a green economy. Finally, school infrastructure and furniture can be built or retrofitted to mitigate environmental impact, and school transportation can also be adjusted to mitigate impact on climate, for instance reducing reliance on fossil fuels. Reviews of empirical work on education and environmental outcomes suggest that while schooling tends to be positively associated with environmental awareness and, in some cases, with behaviors such as recycling or sustainable consumption, evidence on large-scale, durable changes in household energy use or carbon emissions is still limited and methodologically fragile (Ambasz et al., 2025). This reinforces the need to treat these second-order effects not as assumptions, but as empirical questions, and to attend to how curriculum, pedagogy and assessment can jointly support the "knowing, doing and being" dimensions of climate change education (Chang, 2014, 2022; Singh, 2023; Xu & Iyengar, 2023).

Later chapters build on this distinction. Chapter 2 examines systematically what is known about these first- and second-order effects and about the multiple narratives through which education's climate role is conceptualized. Chapters 3–8 analyze, in turn, what students know, how climate is taught, how system levers operate, and how schools connect to other sectors and communities. Here, we introduce the main patterns of first- and second-order effects and illustrate how they are already reshaping education.

1.3.1 First-Order Effects of Climate Change on Education

There are numerous examples of first-order effects as climate change is already disrupting learning in many regions. A recent UNESCO study in Asia Pacific documented how extreme weather events destroyed schools, interrupted schooling as schools were repurposed as emergency shelters and increased dropout rates as a result of climate-induced poverty and displacement. There are now more people displaced by climate disasters than by conflicts (in 2017, 18.8 million new internally displaced people by disasters, compared to 11.8 million displaced due to conflict and violence), and the number of those displaced by climate events is increasing rapidly, from 17.2 million new people displaced by disasters in 2018 to 30.7 million in 2020 (UNESCO, 2023, p. 10).

A study by the United Nations Office for Disaster Risk Reduction shows extensive damage and destruction of school infrastructure resulting from weather-related events: between 2005 and 2017, an average of more than 3200 schools per year were damaged or destroyed in 83 countries, compared to an average of about 412 health facilities per year (UNDRR, 2019). A recent World Bank policy note similarly reports that extreme weather routinely disrupts schooling. For example, Cyclone Freddy in 2023 led to school closures affecting about 5 percent of students in Malawi; in the Philippines, approximately one in five schools is flooded each school year; and the 2022 floods in Pakistan disrupted the education of around 3.5 million children (Venegas Marin et al., 2024). The same report notes that in the period 2005–2024, schools were closed in at least three-quarters of extreme weather events, and that droughts, cyclones and floods frequently lead to prolonged closures when buildings are damaged or used as shelters, as in the 2015 drought in Malawi (when 42% of schools closed) and the 2009 and 2013 cyclones in the Philippines, which damaged thousands of schools and resulted in extended closures (Venegas Marin et al., 2024). The duration of closures is closely linked to the vulnerability and location of school infrastructure and is often longer when schools are used as relocation centers for people displaced by climate hazards; heavy rains also reduce attendance by making travel to school more difficult (Venegas Marin et al., 2024).

In high-income contexts, heat is an increasingly important driver of educational disruption. In the United States, districts have shortened school days or closed schools altogether during heat waves when classrooms lack adequate cooling. For instance, in early September 2023, dozens of schools in Philadelphia implemented abbreviated schedules under the district's Extreme Heat Emergency Response Procedures, and districts in states such as New Jersey and Illinois likewise delayed or shortened school days because of excessive heat (Roston, 2023). During a late-summer heat wave the same year, schools across parts of the Mid-Atlantic and Northeast (including Baltimore, Detroit, Philadelphia and districts in New Jersey and Northeast Ohio) closed or dismissed students early due to dangerously high classroom temperatures (Meckler, 2023). Rising seas and stronger storms also threaten coastal school infrastructure. In Louisiana, for example, hurricanes Katrina and Rita in 2005 damaged or destroyed more than 100 public school buildings in New Orleans alone, displacing tens of thousands of students for months or years (Louisiana State University Libraries, n.d.). Comprehensive reviews of U.S. data estimate that roughly a quarter of public schools are located in census tracts classified as very high-risk for at least one major climate-related hazard, and document growing fiscal pressures on districts from damage, reconstruction and rising insurance premiums (Kraft et al., 2025).

In response to these first-order effects, some schools are installing cooling systems, green roofs, and backup power generators to maintain safe, continuous learning.

After Hurricane Sandy in 2012, the New York City Department of Education developed comprehensive climate resilience guidelines for new school construction and retrofits. In the Philippines, the Department of Education works with municipalities to relocate schools out of high-risk flood and storm surge areas. These adaptation measures help safeguard education as climate hazards increase, demonstrating systemic coordination between education authorities and disaster management. Yet, as Chap. 5 will show, many such efforts remain piecemeal, under-financed and poorly documented, especially in low-income and fragile settings.

1.3.2 Second-Order Effects of Climate Change on Education

As noted above, a key second-order effect of climate change is that it can lead education systems to teach about climate change itself. Around the world, governments and education authorities are increasingly recognizing that education is crucial for building public understanding of climate risks and for enabling participation in mitigation and adaptation efforts (UNESCO, 2021a, 2021b, 2021c, 2021d; UNESCO Global Education Monitoring Report Team & MECCE Project, 2024). This recognition is mirrored in a growing body of research that examines how schooling shapes environmental literacy, attitudes and, more tentatively, behaviors (Ambasz et al., 2025; Angrist, Winseck, Patrinos, & Graff Zivin, 2024). Book-length syntheses such as Chang (2014, 2022) and Xu and Iyengar (2023) conceptualize climate change education as spanning formal and informal settings, emphasize systems thinking, and argue that CCE must connect knowledge, action and identity. Practitioner-oriented guides such as Lê (2025) illustrate how these ideas can be translated into NGSS-aligned climate teaching in secondary science, while investigative work such as Worth (2021) reminds us that climate education can be systematically undermined or misrepresented by political and commercial actors.

In Europe, countries have taken varied approaches to climate change education by reforming curricula and related policies. In Germany, the Federal Ministry of Education and Research has launched a "Climate Education Offensive" (Bildungsoffensive für den Klimaschutz), which aims to strengthen climate and sustainability education through curriculum development, teacher professional learning, and materials that support the energy transition and climate action (Bundesministerium für Bildung und Forschung, 2021; UNESCO Global Education Monitoring Report Team & MECCE Project, 2024). In Italy, Law No. 92 of 20 August 2019 reformed civic education to make climate change and sustainable development a compulsory part of schooling, requiring 33 hours per year of civics education, including climate change, for all students in primary and secondary education (Italy – Climate Change Communication and Education Country Profile, 2023). In England, the national curriculum framework specifies that pupils should learn about "climate and weather" and the impact of human-produced carbon dioxide on climate within science and geography at various key stages, ensuring that all students engage with climate-related content by lower secondary school (Department for Education, 2013; UNESCO Global Education Monitoring Report Team & MECCE Project, 2024). In Sweden, education for sustainable development and sustainable societal planning are cross-cutting themes in the national curriculum, integrated from preschool through upper secondary education and explicitly linked to how human activity affects nature and climate (Swedish National Agency for Education, 2011; Sweden – Climate Change Communication and Education Country Profile, 2023).

Similar trends are visible in other regions. In Africa, many countries are working to align education with climate adaptation and green-economy goals. Kenya's National Environmental Education and Communication Strategy 2018–2022 calls for integrating environmental and climate change content across formal education, non-formal learning and public awareness, emphasizing curriculum development and teacher capacity as central to building climate resilience (National Environment Management Authority, 2018). In South Africa, the National Climate Change Response White Paper explicitly calls for mainstreaming climate change into the education and training system, including technical and vocational education, to support a just transition to a low-carbon, climate-resilient economy (Department of Environmental Affairs, 2011). Regional analyses of climate change communication and education show that a growing number of African countries now reference climate change education in their curricula and sector plans, even though implementation capacity remains uneven (UNESCO, 2023).

In the Asia–Pacific region, several countries have embedded environmental and climate themes in curriculum frameworks and broader education plans. In India, the National Curriculum Framework 2005 identifies environmental education, including issues such as pollution, global warming and sustainable development, as a cross-cutting component to be infused across subjects and grade levels, rather than treated as a stand-alone subject (India – Climate Change Communication and Education Country Profile, 2023; National Council of Educational Research and Training, 2005). In China, the State Council's National Plan for Medium and Long-Term Education Reform and Development (2010–2020) underscores the need to promote "education for the construction of an ecological civilization," encouraging schools and universities to strengthen environmental and sustainability education as part of quality education reform (State Council of China, 2011; UNESCO Global Education Monitoring Report Team & MECCE Project, 2024). In the Philippines, the Department of Education's climate change education initiatives seek to mainstream climate change and disaster risk reduction into basic education through curriculum integration, teacher training and school–community programs (Department of Education – Philippines, n.d.; UNESCO, 2023). Country profiles compiled by UNESCO and the MECCE Project similarly show that Australia and other Asia–Pacific countries have developed national or sectoral strategies encouraging integration of climate change and sustainability topics into schooling, teacher education and community programs (UNESCO Global Education Monitoring Report Team & MECCE Project, 2024).

In the United States, climate change education is highly decentralized. Federal agencies such as NASA and NOAA have long produced curricular materials and data resources on climate science and impacts, but state-level standards and policies vary widely (UNESCO Global Education Monitoring Report Team & MECCE Project, 2024). Some states have begun to move toward more systemic approaches. Maine's Department of Education, for example, has launched a three-phase Climate Education Professional Development Grant Pilot Program, to fund interdisciplinary, climate-related professional learning in partnership with community-based organizations. In its first phase in 2024, the program supported seven projects involving 128 teachers from 50 schools and more than 4000 students; subsequent phases provide several hundred thousand dollars annually for Pre-K–12 professional development and regional "professional learning summits" on climate education (Maine Department of Education, 2024a, 2024b). At the policy level, Illinois became the fifth U.S. state to require climate change education in public schools with the passage of House Bill 4895 (Public Act 103–0837), signed in August 2024. Beginning in the 2026–2027 school year, every public school in the state must provide instruction on climate change, including environmental and ecological impacts and solutions, with the State Board of Education tasked with developing instructional resources and professional learning for educators (Illinois Environmental Council, 2024; Illinois General Assembly, 2024).

These policy developments coexist with what Kraft et al. (2025) describe as a "patchwork" of U.S. PreK–12 responses: some districts and states have articulated climate action and sustainability plans, integrated climate into standards and invested in professional learning, but nationwide policy and practice still fall far short of the scale and urgency of climate risks. Their synthesis underscores that efforts to green school facilities, transportation, food systems and curricula often remain fragmented and under-evaluated, a theme we return to in Chapters 4 and 5. Worth's (2021) journalistic investigation of U.S. classrooms and textbooks further reveals that in some jurisdictions climate change is actively downplayed or misrepresented, producing what she calls "miseducation" rather than education.

Beyond curriculum and policy reforms, a growing landscape of NGOs and community-based initiatives is experimenting with new models of climate education. Comparative analyses document youth-led programs, whole-school transformation projects, and locally grounded climate-action learning in regions ranging from Europe and North America to Africa and Asia, often linked through international networks such as the MECCE Project's case-study program (MECCE Project, 2020; UNESCO Global Education Monitoring Report Team & MECCE Project, 2024). These initiatives treat students as agents of change, emphasize place-based and project-based learning, and help diffuse innovative pedagogies that can complement, and sometimes push, more top-down policy developments. Singh (2023), for example, highlights how stories and justice-centered projects can help learners connect climate science to lived experience and to struggles against inequality.

In addition to teaching and learning, some climate-focused education efforts target school infrastructure and operations. In the United States, for example, the K–12 Climate Action initiative convened by the Aspen Institute developed a national action plan that highlights decarbonizing school buildings and transportation, enhancing resilience of facilities, and leveraging school infrastructure projects as learning opportunities (Aspen Institute & K–12 Climate Action, 2021). At the state level, grant programs like Maine's climate education professional development initiative are often aligned with broader "green schools" and climate-action plans that seek to reduce school-system emissions while integrating climate content into curriculum and career pathways (Maine Department of Education, 2024a, 2024b; National Environment Management Authority, 2018). Kraft et al. (2025) similarly emphasize the importance of approaching these strands together, rather than as isolated projects, so that efforts to reduce schools' environmental impacts, increase resilience, support affected students and staff, strengthen climate curricula and engage communities reinforce one another.

Across regions, these national and subnational initiatives reflect a widening recognition that educating current and future citizens about climate change is essential for building informed public support for climate policy and for equipping learners with the knowledge, skills and dispositions needed for mitigation, adaptation and climate justice (CLEAN, 2025; UNESCO, 2021a, 2021b, 2021c, 2021d, 2023a, 2023b, 2023c, 2023d, 2023e, UNESCO Global Education Monitoring Report Team & MECCE Project, 2024; White et al., 2024). At the same time, reviews of empirical studies show that while education is consistently associated with higher environmental concern, the evidence that general schooling or specific climate education programs produce large, sustained changes in pro-environmental behavior is still limited and uneven (Ambasz et al., 2025; Ardoin et al., 2020; Williamson et al., 2019). While approaches differ according to political priorities and resources, the mainstreaming of climate change education, through curriculum, teacher education, infrastructure and community partnerships, constitutes a critical second-order effect of climate change on education systems worldwide.

The rest of the book deepens this analysis. Chapter 2 maps the dominant narratives in climate change, showing how different perspectives define competencies and pedagogical approaches. These range from climate literacy and education for climate action, to green-skills agendas, ESD, and critical/decolonial approaches. Chapters 5 and 6 explores what climate education looks like in schools and for students themselves.

1.4 Global Education Governance, Green Education and Climate Change Education

Schools have long been described as remarkably stable institutions, "tinkering toward utopia" rather than transforming quickly in response to new demands (Tyack & Cuban, 1995). Yet in little more than a decade, education systems on every continent have begun to acknowledge climate change as a core concern and to reorient, however unevenly, their policies, curricula and partnerships accordingly. This acceleration reflects both the intensification of the climate crisis (IPCC, 2014, 2022a, 2022b) and the consolidation of a global architecture of education governance, centered on the

United Nations and UNESCO, that increasingly treats climate action as intrinsic to the right to education and to education's wider social purposes (UN, 2015; UNESCO, 2020, 2021a). It also reflects growing recognition, in research and policy circles alike, that education is both a sector exposed to climate risks and a potential lever for mitigation and adaptation (Ambasz et al., 2025; Chang, 2014, 2022; Kraft et al., 2025; Sabarwal et al., 2024; Xu & Iyengar, 2023).

Historically, the organization of schooling travelled slowly. Nineteenth-century reformers borrowed ideas across borders, such as John Quincy Adams's admiration for Prussian common schools and Horace Mann's studies of French and German systems, but decades often elapsed between observation and adoption (Tyack & Cuban, 1995). The common school, age-grading, subject-based timetables and normal schools for teacher preparation spread gradually and unevenly as new nation-states consolidated and as public schooling became a marker of modern statehood.

After 1945, that pattern began to change. The creation of the United Nations, the founding of UNESCO and the inclusion of education as a human right in Article 26 of the 1948 Universal Declaration of Human Rights provided a shared normative framework for mass schooling (UN, 1948; UNESCO, 2005). UN-led initiatives such as Education for All and, more recently, Sustainable Development Goal (SDG) 4 further deepened this global architecture, establishing common priorities, shared indicators and regular reporting cycles (UN, 2015; UNESCO, 2020). These platforms turned education into a field of multilateral cooperation and, increasingly, of global policy steering.

It is this evolving architecture of global education governance, nested within the broader UN system of global governance, that has enabled ideas about "green education" and climate change education to spread far more rapidly than earlier reforms (UN, 2015; UNESCO, 2020; UNFCCC, 2015). At the same time, works such as Worth's Miseducation (2021) remind us that these global currents can be resisted or reshaped at national and local levels, producing starkly different climate education experiences within the same global normative environment. Furthermore, emerging ethno-nationalist and right-wing populist movements are challenging post–World War II global governance arrangements not only by seeking more flexible or weaker international environmental agreements, but also by actively contesting multilateral climate norms and obligations (Fu, 2025; Huynh, 2025; Lockwood & Lockwood, 2022). In several regions, these actors frame UN climate and education agendas as elite impositions that threaten national sovereignty, and they work to hollow out or re-nationalize climate governance, for example by diluting national commitments under the Paris Agreement or promoting "climate nationalism" over cooperative solutions. At the level of public opinion and schooling, such movements can blunt the pro-environmental effects of education itself: cross-national evidence suggests that right-wing ideology attenuates the positive relationship between educational attainment and climate change beliefs in more affluent countries (Czarnek et al., 2021), while organized denialist networks linked to conservative and far-right causes have targeted teachers and curricula with climate-misinformation campaigns and materials (Ekwurzel, 2017; Milman, 2023). These dynamics underscore that education systems are embedded in contested political projects and that their potential to advance climate action depends on broader struggles over knowledge, democracy and global cooperation, even as education systems retain significant scope to mitigate their own contributions to climate change through reducing water waste, shifting school food provision away from high-emissions diets, improving energy efficiency in buildings, and promoting more sustainable transportation systems.

1.4.1 From Environmental Education to Education for Sustainable Development and Climate Change Education

The roots of today's climate change education agenda lie in the environmental education movement of the 1970s. UNESCO's Man and the Biosphere (MAB) Program, launched in 1970, sought to build a scientific understanding of human–environment interactions and to link research with education and demonstration sites (UNESCO, n.d.). The 1975 Belgrade Charter and the 1977 Tbilisi Declaration, adopted at UNESCO–UNEP conferences, defined environmental education as a lifelong, holistic process aimed at developing a world population that is aware of and concerned about the environment, and equipped with the knowledge, skills, values and commitment needed to act (UNESCO, 1975, 1977).

A second inflection point came with the 1992 United Nations Conference on Environment and Development in Rio de Janeiro. Agenda 21 called for environmental and development education at all levels and across disciplines (UN, 1992). The United Nations Framework Convention on Climate Change (UNFCCC), adopted at the same conference, established binding commitments on climate action. Education, training, public awareness, public participation and access to information were formalized in Article 6 of the Convention and later elaborated through the Action for Climate Empowerment (ACE) agenda, which explicitly identifies climate change education as one of six pillars of effective climate action (UNESCO, 2017a; UNFCCC, 1992, 2015).

Throughout the 2000s, UNESCO's leadership of the United Nations Decade of Education for Sustainable Development (DESD, 2005–2014) shifted the focus from discrete environmental topics toward "education for sustainable development" (ESD): a broader vision that connects environmental integrity, social justice and economic viability (UNESCO, 2005, 2010b, 2014a). The Berlin Declaration on ESD, adopted at UNESCO's 2021 World Conference, urged countries to make climate action "a core curriculum component" by 2025 and to embed ESD's cognitive, socio-emotional and action-oriented dimensions across systems (UNESCO, 2021b). UNESCO's current ESD for 2030 framework consolidates these shifts. It presents ESD as a lifelong process that equips learners with the knowledge, skills, values and attitudes needed to tackle climate change, biodiversity loss, overuse of resources and inequality, while emphasizing transformative pedagogies and whole-institution approaches (UNESCO, 2020). Within this framework, climate change education is treated not as a niche specialization but as a central strand of ESD (UNESCO, 2017a, 2020).

Within this broader evolution, climate change education has increasingly been conceptualized as a specific, though closely related, field. Beyond UNESCO's policy documents, scholarly work has helped to delineate its scope and emphases. In geography and environmental education, Chang's Climate Change Education: Knowing, Doing and Being (2014, 2022) defines CCE for the "global citizen," analyzes how it appears in formal curricula, examines teacher readiness and assessment, and argues that robust climate change education must simultaneously develop students' understanding ("knowing"), their capacity to act ("doing") and their identities and values as environmentally responsible citizens ("being"). Singh's Teaching Climate Change: Science, Stories, Justice (2023) develops a complementary, justice-centered, transdisciplinary pedagogy that combines physical science, stories and attention to climate emotions. Xu and Iyengar's (2023) primer, written from a sustainability-science perspective, emphasizes systems thinking across sectors and settings, foregrounding structural inequalities and "future workforces." Together with practice-oriented guides such as Lê (2025), which works at the scale of secondary science classrooms, and investigative accounts such as Worth (2021), which expose how climate education is distorted in some contexts, these books exemplify how CCE is emerging as a field at the intersection of policy, research and practice.

1.4.2 Climate Treaties and the Normative Anchoring of Climate Change Education

Over the last decade, climate agreements have entrenched education as a formal obligation of states. The UNFCCC's ACE agenda calls on governments to develop educational and public awareness programs, train personnel, foster public participation and cooperation, and ensure access to information on climate change (UNFCCC, 1992, 2012). The 2015 Paris Agreement reaffirmed this by dedicating Article 12 to "enhanc[ing] climate change education, training, public awareness, public participation and public access to information," and by linking these measures explicitly to the effectiveness of mitigation and adaptation (UNFCCC, 2015).

These legal commitments are echoed in SDG 13, whose target 13.3 urges all countries to "improve education, awareness-raising and human and institutional capacity on climate change mitigation, adaptation, impact reduction and early warning," and in SDG 4.7, which calls for all learners to acquire knowledge and skills needed to promote sustainable development (UN, 2015). Education is thus inscribed both as a means of climate action and as an object of climate resilience.

UN agencies have increasingly framed education as a "critical agent" in confronting the climate crisis. A 2020 UN climate brief, for example, argues that education can encourage people to change attitudes and behavior, help them make informed decisions and empower especially young people to take action, reducing the fear that often surrounds climate debates (UN, 2020).

UNESCO's own definition captures this dual role: climate change education "helps people understand and address the impacts of the climate crisis, empowering them with the knowledge, skills, values and attitudes needed to act as agents of change" (UNESCO, n.d.). In short, international law, global development goals and UN policy discourse now converge on the idea that quality education is both threatened by climate change and indispensable to any just and effective response (UN, 2015; UNESCO, 2017a, 2020, 2021a; UNFCCC, 2015).

1.4.3 UNESCO's Contribution: From Ideas to Standards, Capacity and Coalitions

Within this architecture, UNESCO has used the full range of global education governance functions, knowledge production, standard-setting, capacity building and coalition-building, to advance what it variously terms environmental education, education for sustainable development and climate change education.

(a) Generating and disseminating ideas

UNESCO's climate change education work is anchored in its Climate Change Education for Sustainable Development program and in the broader ESD for 2030 agenda, which together aim to raise climate literacy and to support learners of all ages in developing the knowledge, skills, values and agency needed for mitigation, adaptation and climate justice (UNESCO, 2017a, 2020, 2023a).

Recent analytical work has sharpened this agenda. The 2024 Global Education Monitoring (GEM) paper Education and climate change: Learning to act for people and planet, produced with the Monitoring and Evaluating Climate Communication and Education (MECCE) Project, argues that education's climate potential remains underused and that climate change education needs to move decisively beyond information transmission towards socio-emotional, ethical and action-oriented learning (UNESCO, 2024a). Based on a global analysis of curriculum documents, the paper reports that 76 countries score, on average, only about half of the possible points on environment and sustainability content, and much lower on climate change and biodiversity. "Green" content is more common in science than in social science syllabi, and more prevalent in upper than lower grades; evidence that early and cross-curricular engagement with climate issues remains limited (Benavot & McKenzie, 2026;McKenzie & Benavot, 2024; UNESCO, 2024a).

Complementing this, the GEM team and MECCE have produced climate change communication and education country profiles for around 80 countries, documenting the extent to which climate change is reflected in national education laws, policies, curricula and teacher education (UNESCO, 2023a; UNESCO Global Education Monitoring Report Team & MECCE Project, 2024). These profiles make visible the large proportion of countries that lack specific legislation or budgets for climate change education and the gaps between policy statements and implementation.

UNESCO has also drawn attention to climate-related threats to the right to education. A recent global report on climate change, displacement and education synthesizes evidence on how disasters and slow-onset climate impacts are driving internal displacement, disrupting schooling and exacerbating educational inequalities, and offers guidance to governments on protecting the right to education of climate-displaced populations (UNESCO, 2023c; UNESCO; United Nations University Institute for the Advanced Study of Sustainability; & UNESCO Bangkok, 2023).

Finally, UNESCO's report Reimagining Our Futures Together: A New Social Contract for Education, prepared by an independent expert commission on the Futures of Education, situates climate change, biodiversity loss and widening inequality as intertwined, existential crises. It documents how extreme weather events and environmental degradation already damage schools, interrupt learning and threaten the right to education, especially for marginalized communities, and calls for a radical rethinking of education's aims, content and organization in light of planetary boundaries (IPCC, 2022a, 2022b; UNESCO, 2021d).

(b) Acting as a clearinghouse and evidence broker

As a clearinghouse, UNESCO collates and disseminates evidence on climate change education policies and practices. The report, Youth Demands for Quality Climate Change Education, based on surveys of more than 17,000 young people in 166 countries, reveals that 70% feel they cannot explain climate change well and many judge their school climate education to be superficial and passive (UNESCO, 2022).

The GEM/MECCE country profiles, noted above, extend this work by systematically tracking whether and how climate change is integrated into national policies, curricula, teacher training and assessment, and by highlighting the scarcity of dedicated budgets for climate communication and education (UNESCO, 2023a, 2024a).

Within UNESCO's publishing program, curriculum studies such as Climate Change and Sustainability in Science and Social Science Secondary School Curricula (covering 85 countries) and the companion primary-level study (88 countries) provide detailed baselines on the presence and depth of climate and sustainability content in science and social science syllabi (Benavot & McKenzie, 2026; McKenzie & Benavot, 2024). They show that majorities of curricula either omit or give only low to moderate attention to climate change, especially in early grades and in social science subjects, underscoring how far most systems remain from "climate-ready" curricula (McKenzie & Benavot, 2024; UNESCO, 2024a).

(c) Norm- and standard-setting

UNESCO's norm-setting role is visible both in broad ESD frameworks and in climate-specific standards. The Berlin Declaration on ESD (2021) urged all countries to integrate ESD into education systems at all levels and specifically called for climate action to be treated as a core curriculum component by 2025 (UNESCO, 2021b).

Building on these commitments, UNESCO and partners launched the Greening Education Partnership (GEP) at the 2022 Transforming Education Summit, with the mission of "getting every learner climate-ready." GEP articulates four pillars, greening schools, greening curriculum, greening teacher training and system capacities, and greening communities, and sets ambitious global targets, such as doubling the number of countries that include climate education in curricula and greening at least half of all schools by 2030 (UNESCO, 2023b, 2024c).

To operationalize these goals, UNESCO has developed new global guidance. Greening Every Curriculum: Guidance to Teach and Learn for Climate Action (2024a) proposes a common language and age-progressive learning outcomes for climate and sustainability education across cognitive, socio-emotional and behavioral domains, emphasizing action-orientation, justice, scientific robustness and lifelong learning. The companion Green School Quality Standard: Greening Every Learning Environment (UNESCO, 2024b) specifies criteria across four domains, governance, facilities and operations, teaching and learning, and community engagement, to guide ministries, accreditation bodies and schools in designing and recognizing climate-ready institutions.

At COP28 in 2023, UNESCO worked with the COP Presidency and countries to launch the Declaration on Education and Climate Change, focused on adaptation, mitigation and investment. Thirty-nine countries endorsed the declaration, committing to climate-resilient education systems, stronger integration of climate content in curricula and teacher training, and greater financing for climate-related education measures (UNESCO, 2023b).

(d) Building capacity and providing technical cooperation

Through GEP and related initiatives, UNESCO provides technical support to ministries and other actors seeking to "green" their systems. GEP's latest figures show that over 80 countries and more than 1100 organizations have joined the partnership; among participating countries, roughly three-quarters plan to revise curricula to integrate climate change and biodiversity, and a similar proportion plan dedicated climate training for teachers in the next few years (UNESCO, 2023b, 2024c).

UNESCO's International Bureau of Education (IBE) has made curriculum transformation for sustainability and climate action a flagship priority, supporting countries to embed climate and sustainability across subjects and levels, align curricula with SDG 4.7 and climate objectives and adopt transformative pedagogies that connect cognitive, socio-emotional, behavioral and ethical dimensions of learning (UNESCO International Bureau of Education, 2016, 2018, 2022, 2023a, 2023b, 2024a, 2024b).

UNESCO and partners also invest in teacher professional development, open educational resources and regional capacity-building, for example, through work on "greening TVET" and on green skills for just transitions (UNESCO–UNEVOC, 2021).

(e) Catalyzing international cooperation and public engagement

UNESCO has been instrumental in keeping education visible in UNFCCC processes. At successive Conferences of the Parties (COP27, COP28 and on the road to COP29), it has convened ministerial roundtables, youth dialogues and technical sessions on climate change education for social transformation, often in partnership with UNFCCC, UNICEF, the World Bank and others (UNESCO, 2023a, 2023b; UNFCCC, 2012). COP28 saw the first-ever education pavilion, co-hosted by the UAE Ministry of Education and UNESCO, and a dedicated Greening Education Hub with more than 200 side events (UNESCO, 2023b).

Beyond formal negotiations, UNESCO collaborates with actors such as the World Meteorological Organization (WMO) and Climate Central on public communication campaigns. Initiatives like WMO and Climate Central's Summer in the Cities and earlier "weather reports from 2050" video series bring climate projections to life through familiar broadcast formats, illustrating how climate change will shape local weather in cities worldwide and providing compelling resources for classrooms and community discussion (WMO, 2014; WMO & Climate Central, 2017).

UNESCO's #GreeningEducation campaign and its global "Conversations on climate change education for social transformation" further extend this catalytic role, mobilizing students, teachers, schools and communities in the lead-up to COPs and amplifying youth demands for high-quality climate education (UNESCO, 2022, 2023a, 2023d).

1.4.4 The IBE–UNESCO "Green Education for Sustainable Societies" Lens

Within this broader landscape, the IBE's Green Education for Sustainable Societies (GESS) initiative offers a useful lens for thinking about system-wide responses to climate change through education. Building on UNESCO's Learn for our planet review, the ESD for 2030 roadmap and emerging empirical evidence on effective climate pedagogy, GESS rests on two premises (Iyengar & Kwauk, 2021; UNESCO, 2020, 2021a, 2024a).

First, green education must go beyond token references to "the environment" or "climate" to become holistic, transdisciplinary and justice-oriented. This implies curricula that integrate cognitive, socio-emotional, behavioral and ethical dimensions; link local and global perspectives; value Indigenous and local knowledge; and attend to issues of gender, power and intergenerational equity (IPCC, 2014, 2022a, 2022b; Kwauk & Casey, 2021; Leicht et al., 2018; Rousell & Cutter-Mackenzie-Knowles, 2020). Research shows, for instance, that intensive, high-quality climate courses can reduce learners' lifetime carbon emissions at a scale comparable to some technological mitigation measures (Cordero et al., 2020) and that child-led, place-based climate education can influence parents' attitudes and behaviors (Lawson et al.,

2019). Systematic reviews likewise highlight the importance of participatory, experiential, arts-based and interdisciplinary approaches over purely didactic methods (Adlit & Adlit, 2022; Monroe et al., 2019; Rousell & Cutter-Mackenzie-Knowles, 2020). Ambasz et al.'s (2025) review, which focuses on general schooling rather than specialized climate curricula, converges with these findings in emphasizing that knowledge gains alone rarely suffice; affective engagement, social norms and enabling conditions matter for sustained pro-environmental behavior.

Second, curriculum does not exist in a vacuum. What is written in syllabi interacts with governance arrangements, assessment regimes, teacher education, school culture, community learning ecologies and the physical conditions of educational infrastructure (Iyengar & Kwauk, 2021a, 2021b; UNESCO International Bureau of Education, 2016, 2018, 2022, 2023a, 2024a, 2024b). This system-level view is echoed in national-level syntheses that map how schools' environmental impacts, resilience, supports for affected communities, instructional practices and broader societal roles intersect (Kraft et al., 2025).

On this basis, GESS conceptualizes green education along four inter-connected dimensions that we will revisit throughout this book (UNESCO International Bureau of Education, 2022, 2023b):

- Inclusive and participatory governance. Green education requires building coalitions among governments, youth, educators, civil society and the private sector; aligning education plans with broader climate and development strategies; and recognizing young people as co-agents in setting agendas rather than passive recipients of information.
- Transformative teaching, learning and assessment. Climate-relevant curricula should be interdisciplinary and competency-based, combining foundational climate and systems knowledge with socio-emotional learning, ethical reflection and opportunities for individual and collective action. Teachers need sustained professional development and collaborative spaces to co-design and enact such curricula (Iyengar & Kwauk, 2021a, 2021b; UNESCO International Bureau of Education, 2016, 2022).
- Communitarian and lifelong learning ecosystems. Green education unfolds across formal, non-formal and informal settings, schools, communities, workplaces, digital spaces and social movements. Evidence from intergenerational learning studies shows that children can influence household and community practices, suggesting that education systems should see themselves as hubs in broader green learning ecosystems (Adlit & Adlit, 2022; Lawson et al., 2019; Mbah et al., 2022; Kraft et al., 2025).
- Sustainable and locally pertinent infrastructure. School buildings, grounds and operations can serve both as symbols and laboratories of climate action. "Whole-school" approaches that align physical infrastructure, resource use, governance, pedagogy and community engagement can model sustainable lifestyles and enhance resilience to climate shocks, an agenda now being concretized through UNESCO's Green School Quality Standard (UNESCO, 2024b; UNESCO International Bureau of Education, 2024b).

These four dimensions provide a framework for analyzing how far national systems are moving toward climate-ready education, and where gaps remain, in later chapters of this book.

1.4.5 Other Multilateral and Regional Actors

UNESCO is not alone among the organizations that form the global governance education architecture in advancing education for climate action. Other UN agencies, multilateral banks and regional bodies have begun to treat climate change education as both a risk management imperative and a development opportunity.

The World Bank's 2024 flagship report Choosing Our Future: Education for Climate Action synthesizes emerging evidence on the bidirectional relationship between climate change and education. It estimates that hundreds of millions of students have already experienced climate-related school closures, with low-income countries hardest hit (Sabarwal et al., 2024). The report argues that education is "one of the most powerful yet underutilized tools for climate action": educational attainment is the strongest predictor of climate awareness; quality schooling can foster mitigation and adaptation behaviors; and broad-based green skills are essential for a just transition. Yet education receives less than 2% of climate finance, and climate risks are rarely integrated into education sector plans (Sabarwal et al., 2024).

The report proposes a policy agenda that overlaps strongly with the GEP and GESS frameworks: integrating locally relevant climate content and problem-solving into curricula; strengthening teachers' capacity for climate pedagogy; investing in climate-resilient, thermally comfortable school infrastructure; and equipping learners with technical, cognitive and socio-emotional "green skills" for both new and existing jobs (Sabarwal et al., 2024; UNESCO–UNEVOC, 2021).

The OECD's Trends Shaping Education 2025 likewise places climate change and ecological crises at the center of its analysis of megatrends. One chapter, "Education on a fragile planet," highlights how climate change, biodiversity loss and pollution simultaneously threaten education systems, through damaged infrastructure, health risks and learning disruptions, and increase demand for education that builds environmental literacy, supports sustainable lifestyles and prepares learners for green jobs (OECD, 2025). The report calls on education systems to strengthen interdisciplinary, place-based and futures-oriented learning, foster systems thinking and active citizenship, and address climate misinformation.

Regional research is beginning to examine how these global agendas translate into national policies and local practices. Mbah et al. (2022) analysis of climate change education policies in South Asia, for example, documents how climate change education is now referenced in many national climate strategies, education plans and public awareness programs across Afghanistan, Bangladesh, Bhutan, India, Maldives, Nepal, Pakistan and Sri Lanka. Yet significant gaps remain between policy aspirations and implementation: CCE is often fragmented across sectors, under-resourced and weakly connected to local adaptation needs. The authors propose a support framework that combines theory of change, critical pedagogy and place-based approaches to help bridge these gaps and make climate change education a lever for equitable adaptation (Mbah et al., 2022).

At the same time, a growing body of educational research is probing what effective climate change education looks like in classrooms and communities. Adlit and Adlit's (2022) qualitative systematic review of studies published between 2019 and 2021 identifies promising approaches (participatory and inquiry-based pedagogies, interdisciplinary projects, place-based and socio-cultural learning, arts and humanities integration, digital simulations and games) as well as recurring constraints such as overloaded curricula, limited teacher preparation and the emotional weight of climate learning for both students and teachers (see also Monroe et al., 2019; Rousell & Cutter-Mackenzie-Knowles, 2020). These findings resonate with, and empirically underpin, UNESCO's calls for climate change education that is engaging, transformative and attentive to learners' emotional realities (UNESCO, 2017a, 2020, 2023a, 2024a). Ambasz et al. (2025) add a complementary perspective by reviewing studies that link general educational attainment, not only explicit climate education, to environmental attitudes and behaviors, and by arguing for more quasi-experimental and longitudinal work to illuminate these links.

1.4.6 From Global Norms to Classroom Realities

Taken together, these multilateral agendas, normative instruments and research syntheses signal that climate change education is no longer a peripheral concern. It is embedded in core human-rights treaties, climate agreements, SDG targets, UN programs and donor strategies (Ambasz et al., 2025; OECD, 2025; Sabarwal et al., 2024; UN, 2015; UNESCO, 2017a, 2020, 2021a; UNFCCC, 2015). Yet, the depth and quality of implementation remain highly uneven.

UNESCO's own analyses show that nearly half of national curriculum frameworks still make no reference to climate change, that only a minority of teachers feel confident teaching climate action and that young people frequently judge their climate education to be shallow or disconnected from action (UNESCO, 2022, 2023a, 2024a). World Bank and OECD reports underline that education is both acutely vulnerable to climate shocks and dramatically under-represented in climate finance and policy (OECD, 2025; Sabarwal et al., 2024). Empirical studies in regions such as South Asia, and systematic reviews of classroom practice worldwide, reveal policy–practice gaps, persistent inequalities and the need for context-sensitive, justice-oriented approaches (Adlit & Adlit, 2022; Mbah et al., 2022; Monroe et al., 2019; Rousell & Cutter-Mackenzie-Knowles, 2020). National-level syntheses, such as that of Kraft et al. (2025) for the U.S., further highlight how fragmented responsibilities, fiscal constraints and limited data impede coherent climate strategies for school systems. Investigative journalism such as Worth (2021) shows how, even where national and global norms favor strong climate education, local politics, textbook markets and organized interests may actively undermine it.

At the same time, reviews of the education–environment literature emphasize how much remains unknown about the causal impact of schooling and specific educational interventions on environmental behaviors and outcomes (Ambasz et al., 2025). Xu and Iyengar (2023) similarly call for more attention to inequalities of access to climate education and to the ways in which systems thinking can inform evaluation and design.

The remainder of this book asks how far these global aspirations have filtered into national systems, schools and classrooms, and what forms of systemic change would be required for climate-ready education to become the norm rather than the exception. Chapters 4–8, in particular, return to the four GESS dimensions, governance, pedagogy and assessment, lifelong learning ecosystems and infrastructure, to examine how different countries are, or are not, mobilizing curricula, teacher development, community partnerships and educational spaces in response to climate change.

1.5 Structure and Approach of the Book

The remainder of the book moves from concepts and evidence to practice and policy, while keeping in view the systems, timescales and governance perspectives introduced in this chapter.

Chapter 2 maps the terrain of perspectives on education and climate change, elaborating the distinction between first-order effects (how climate disrupts education) and second-order effects (how education systems respond and act on climate change). It synthesizes evidence on the ways climate hazards affect learning, infrastructure, health and life-course trajectories, and reviews emerging strategies for educational continuity and resilience. It then identifies and analyzes five coexisting narratives about education's role in climate change: climate literacy, education for climate action, education for the green economy, education for sustainable development, and critical/decolonial climate education, examining their assumptions, overlaps and tensions, and relating them to the frameworks proposed by Chang (2014, 2022), Singh (2023), Xu and Iyengar (2023), Lê (2025) and others.

Chapter 3 turns to national and subnational systems. Using detailed MECCE country profiles and other policy sources, it analyzes how governments in Italy, Scotland, Aotearoa New Zealand, the Dominican Republic, the United States and Mexico have framed education's climate role in law and strategy, and through which instruments, curriculum, teacher education, infrastructure, assessment, partnerships, they have tried to operationalize these ambitions. The chapter draws out cross-case patterns in governance, finance and implementation, and identifies the constraints that keep many climate-education efforts at the level of pilots or rhetoric.

Chapter 4 examines the role of educator and civil society networks, teacher professional communities, school alliances, NGOs, faith-based and private-school networks, and digital platforms, in advancing climate literacy and practice. It considers how bottom-up innovations emerge and spread, how professional learning communities and online networks support teachers where formal guidance is weak, and how transnational initiatives (for example, Eco-Schools, the Green Schools Alliance, Teach for All, Ignatian and Laudato Si'-inspired school networks) function as de facto infrastructures for climate education. The chapter relates these dynamics to Cohen and Mehta's (2017) work on "niche" reforms and their potential to inform and pressure wider system change.

Chapter 5 explores how climate change is taught in practice, across classrooms and communities. Drawing on global policy reviews, a set of 11 in-depth MECCE case studies of approaches to climate change in various countries, and 39 exemplary schools advancing climate education from all world regions, it highlights the gaps between national curriculum frameworks and what is enacted, examines subject-specific and cross-curricular approaches, and presents exemplars of humanities-based, STEM-integrated, place-based, Indigenous and whole-school climate education.

Chapter 6 investigates what students around the world actually know about climate change, what they believe and feel, and how they act with respect to climate. It synthesizes findings from large-scale assessments (PISA, TIMSS, PIRLS, ICCS), recent systematic reviews of student conceptions and misconceptions, and meta-analyses of climate-education interventions. It then compares existing climate-literacy and sustainability-competence frameworks (including IPCC-informed models, UNESCO's ESD learning objectives and regional approaches) and proposes a synthesized framework for "climate literacy as agency," integrating scientific understanding, systems thinking, socio-emotional capacities and action dispositions.

Chapter 7 proposes a systems approach to advance effective climate education policy through the coherent alignment of curriculum, assessment, teacher preparation and professional development, school leadership, and governance. Building on Reimers' (2020a, 2020b) five perspectives on educational change (cultural, psychological, professional, institutional and political) and frameworks of policy coherence it examines how climate-relevant competencies can be coherently embedded in standards, assessments, professional learning and school-improvement processes. It integrates recent syntheses on curriculum integration, teacher professional development and school leadership in climate-change education to show how systems drift into "low-climate-learning traps" and what it would take to escape them.

Chapter 8 expands the focus to the intersection of education with other sectors and the broader ecological and urban context, treating schools as social, physical and cultural anchors for climate resilience. It draws on examples such as Reggio Emilia-inspired municipal approaches, "learning cities," community schools and initiatives like the Aspen Institute's K–12 Climate Action plan to explore how schools connect with health, social protection, urban planning, disaster-risk management and local climate-action planning. In line with research agendas proposed by Kraft et al. (2025) and others, it pays particular attention to schools' roles as community hubs for climate awareness, care and coordinated response during environmental crises.

Chapter 9 concludes by synthesizing key findings across the book, identifying gaps and priorities for future work, and offering actionable recommendations for leveraging education systems as engines of climate adaptation, mitigation and transformation. It revisits the systems, timescales and governance perspectives introduced here, reflects on how the five narratives of climate change education play out in practice, and sketches what "green education for sustainable societies" might mean in diverse contexts.

Throughout, our approach is comparative and mixed-method. We draw on:

Literature review and synthesis across climate science, environmental education, education for sustainable development, development economics and comparative education. Comparative policy and curriculum analysis, especially of how climate and environmental issues figure in national education sector plans and curriculum frameworks, building on UNESCO's Learn for our planet and related reviews and country profiles (UNESCO, 2018, 2019; UNESCO Global Education Monitoring Report Team & MECCE Project, 2023a–2023f, 2024; UNESCO International Bureau of Education, 2016, 2018, 2022).

Case studies of schools, systems and networks that exemplify particular approaches to climate education, including the 11 MECCE case studies, the 39 T4Education "Best School" finalists, Reggio Emilia-inspired municipal systems, and whole-school and community-based models (Atanga et al., 2025; Finnegan et al., 2025; Golshani et al., 2025; Kagawa & Selby, 2010, 2012; Karsgaard et al., 2025; Kurup et al., 2025; Sabbatini & Beech, 2025; Shepardson et al., 2009, 2011, 2012).

Systematic and scoping reviews on specific levers, including recent syntheses of curriculum integration, teacher professional development and school leadership in climate-change education.

Systems-oriented analysis of how education and climate systems interact, drawing on complexity science and on frameworks for understanding educational change as simultaneously cultural, psychological, professional, institutional and political (Mitchell, 2009; Meadows, 2008; Reimers, 2020a, 2020b, 2021; World Bank, 2018).

Where direct evidence on climate change education is limited, we draw on adjacent fields such as disaster-risk-reduction education, health education and global citizenship education to identify relevant lessons (Kagawa & Selby, 2010, 2012; Shepardson et al., 2011; UNESCO, 2018, 2019). Our analytic lens is explicitly systemic and equity-oriented: we are attentive to how climate change both reflects and reinforces existing inequalities, and to how education responses may either mitigate or reproduce those patterns. We also situate our work in relation to emerging research agendas that call for more rigorous, policy-relevant studies on how schooling affects environmental outcomes and how schools can reduce their own environmental impacts while enhancing resilience and community well-being (Ambasz et al., 2025; Kraft et al., 2025; Xu & Iyengar, 2023).

In this way, the book aims not only to describe the current landscape of education and climate change, but also to contribute to ongoing efforts as well as national research agendas to reorient education systems toward sustainable, healthy and inclusive futures.

References

Adlit, M. F., & Adlit, M. F. (2022). Teaching climate change: A systematic review from 2019–2021. *International Journal of Advanced Research and Publications, 5*(5), 12–19.

Ambasz, D., Gupta, A., & Patrinos, H. A. (2025). A review of human development and environmental outcomes. *Economics Bulletin, 45*(4), 1687–1709.

Ardoin, N. M., Bowers, A. W., & Gaillard, E. (2020). Environmental education outcomes for conservation: A systematic review. *Biological Conservation, 241*, Article 108224. https://doi.org/10.1016/j.biocon.2019.108224

Aspen Institute, & K–12 Climate Action. (2021). K–12 climate action: A policy roadmap for education leaders. Aspen Institute.

Bar-Yam, Y. (1997). *Dynamics of complex systems*. Perseus Books/Addison Wesley.

Benavot, A., & McKenzie, M. (2026). *Climate change and sustainability in science and social science primary school curricula*. UNESCO.

Bundesministerium für Bildung und Forschung. (2021). *Bildungsoffensive für den Klimaschutz* [Climate education offensive]. Federal Ministry of Education and Research.

CLEAN. (2025). *Education pathways to climate outcomes and their potential for climate finance*. Report prepared by C. Kwauk for the Unbounded Alliance for the Climate, Environment, and Nature (CLEAN) Helpdesk. https://www.clean-helpdesk.org/

Cohen, D. K., & Ball, D. L. (1999). *Instruction, capacity, and improvement* (CPRE Research Report Series RR 43). Consortium for Policy Research in Education.

Cohen, D. K., & Mehta, J. D. (2017). Why reform sometimes succeeds: Understanding the conditions that produce reforms that last. *American Educational Research Journal, 54*(4), 644–690. https://doi.org/10.3102/0002831217700078

Cordero, E. C., Centeno, D., & Todd, A. M. (2020). The role of climate change education on individual lifetime carbon emissions. *PLoS ONE, 15*(2), Article e0206266.

Chang, C. H. (2014). *Climate change education: Knowing, doing and being*. Routledge.

Chang, C. H. (2022). *Climate change education: Knowing, doing and being* (2nd ed.). Routledge.

Chang, C. H., & Pascua, L. (2015). Singapore students' misconceptions of climate change. *International Research in Geographical and Environmental Education, 25*(1), 84–96. https://doi.org/10.1080/10382046.2015.1106206

Czarnek, G., Kossowska, M., & Szwed, P. (2021). Right-wing ideology reduces the effects of education on climate change beliefs in more developed countries. *Nature Climate Change, 11*(1), 9–13. https://doi.org/10.1038/s41558-020-00930-6

Darling-Hammond, L. (2010). *The flat world and education: How America's commitment to equity will determine our future*. Teachers College Press.

Department for Education. (2013). *The national curriculum in England: Framework document*. Department for Education.

Department of Education – Philippines. (n.d.). *Climate change education in basic education*. Department of Education.

Department of Environmental Affairs. (2011). *National climate change response white paper*. Government of South Africa.

Ekwurzel, B. (2017, April 7). Déjà vu all over again: Heartland Institute peddling misinformation to teachers about climate change. Union of Concerned Scientists Blog. https://blog.ucsusa.org/brenda-ekwurzel/deja-vu-all-over-again-heartland-institute-peddling-misinformation-to-teachers-about-climate-change/

Elmore, R. F. (1996). Getting to scale with good educational practice. *Harvard Educational Review, 66*(1), 1–26.

Finnegan, W., Drayson, R., Guruprasad, S., & Runkle, Q. (2025). *Empowering young people to teach their teachers: Discovering quality climate education through digital storytelling*. MECCE Project Funded Case Study.

Fu, C. (2025). The populistisation of green politics and the greening of populism: An examination of environmental populism in Europe. *International Journal of Anthropology and Ethnology, 9*(1), 1–24. https://doi.org/10.1186/s41257-024-00124-4

Golshani, Z., Masoumi, S., Chalabiyani, B., & Alavijeh, E. E. (2025). *A student-centred climate education program for high school girls in Iran.* MECCE Project Funded Case Study.

Hestness, E., McDonald, R. C., Breslyn, W., McGinnis, J. R., & Mouza, C. (2014). Science teacher professional development in climate change education informed by the Next Generation Science Standards. *Journal of Geoscience Education, 62*(3), 319–329. https://doi.org/10.5408/13-049.1

Holland, J. H. (1998). *Emergence: From chaos to order*. Perseus Books.

Howe, P. D., & Leiserowitz, A. (2013). Who remembers a hot summer or a cold winter? The asymmetric effect of beliefs about global warming on perceptions of local climate conditions in the U.S. *Global Environmental Change*, 23(6), 1488–1500. https://doi.org/10.1016/j.gloenvcha.2013.09.014

Hulme, M. (2014). Attributing weather extremes to "climate change": A review. *Climatic Change, 127*(2), 325–338.

Huynh, N. K. (2025). Right-wing populists and the politics of renegotiating international environmental agreements. *Global Environmental Politics, 25*(1), 21–45. https://doi.org/10.1162/glep_a_00770

Illinois Environmental Council. (2024, August 12). Illinois becomes fifth state to require climate change education in public schools.

Illinois General Assembly. (2024). Public Act 103-0837 (House Bill 4895): An act concerning climate change education in public schools.

India – Climate Change Communication and Education Country Profile. (2023). *In Climate change communication and education country profiles: Approaches to greening education around the world.* UNESCO Global Education Monitoring Report/MECCE Project.

Intergovernmental Panel on Climate Change. (2001). Climate change 2001: The scientific basis. In *Contribution of Working Group I to the third assessment report of the intergovernmental panel on climate change.* Cambridge University Press.

Intergovernmental Panel on Climate Change. (2014). Climate change 2014: Synthesis report. In *Contribution of Working Groups I, II and III to the fifth assessment report of the intergovernmental panel on climate change.* IPCC.

Intergovernmental Panel on Climate Change. (2018). *Global warming of 1.5 °C: An IPCC special report on the impacts of global warming of 1.5 °C above pre-industrial levels and related global greenhouse gas emission pathways, in the context of strengthening the global response to the threat of climate change, sustainable development, and efforts to eradicate poverty.* IPCC.

Intergovernmental Panel on Climate Change. (2022a). Climate change 2022: Impacts, adaptation, and vulnerability. In *Contribution of Working Group II to the sixth assessment report of the intergovernmental panel on climate change.* Cambridge University Press.

Intergovernmental Panel on Climate Change. (2022b). Climate change 2022: Mitigation of climate change. In *Contribution of Working Group III to the sixth assessment report of the intergovernmental panel on climate change.* Cambridge University Press.

Italy – Climate Change Communication and Education Country Profile. (2023). *In Climate change communication and education country profiles: Approaches to greening education around the world.* UNESCO Global Education Monitoring Report/MECCE Project.

Iyengar, R., & Kwauk, C. (2021). *Education for climate action in the context of climate justice.* Brookings Institution.

Iyengar, R., & Kwauk, C. T. (Eds.). (2021). *Curriculum and learning for climate action: Toward an SDG 4.7 roadmap for systems change.* Brill|Sense/UNESCO IBE.

Kagawa, F., & Selby, D. (2010). Ready for the storm: Education for disaster risk reduction and climate change adaptation and mitigation. *Journal of Environmental Education, 41*(4), 217–227. https://doi.org/10.1080/00958960903227082

Kagawa, F., & Selby, D. (2012). Ready for the storm: Education for disaster risk reduction and climate change adaptation and mitigation. *Journal of Environmental Education, 41*(4), 217–227. https://doi.org/10.1080/00958964.2010.505415
Karsgaard, C., Michel, L. J., & Catholique, I. (2025). Climate change and futures-building through Dene stewardship camps. MECCE Project Funded Case Study.
Komatsu, H., & Rappleye, J. (2020). Will education post-2015 move us toward environmental sustainability? In A. Wulff (Ed.), *Grading goal four: Tensions, threats, and opportunities in the Sustainable Development Goal on quality education* (pp. 297–321). Brill. https://doi.org/10.1163/9789004430365_014
Kraft, M. A., Malik, S., & Falken, G. T. (2025, July 22). *Education and climate change: Synthesizing the evidence to guide future research* [Unpublished manuscript]. Brown University.
Kwauk, C., & Casey, O. (2021). *A new green learning agenda: Approaches to quality education for climate action*. Brookings Institution.
Lawson, D. F., Stevenson, K. T., Peterson, M. N., Carrier, S. J., Strnad, R. L., & Seekamp, E. (2019). Children can foster climate change concern among their parents. *Nature Climate Change, 9*, 458–462.
Leicht, A., Heiss, J., & Byun, W. J. (Eds.). (2018). *Issues and trends in education for sustainable development*. UNESCO Publishing.
Lockwood, B., & Lockwood, M. (2022). How do right-wing populist parties influence climate and renewable energy policies? Evidence from OECD countries. *Global Environmental Politics, 22*(3), 12–37. https://doi.org/10.1162/glep_a_00659
Louisiana State University Libraries. (n.d.). Impact on education – Louisiana hurricanes: Hurricane Katrina impact on education.
Maine Department of Education. (2024a, March 13). *Maine DOE releases professional development grant for 2024/2025 school year to promote climate education in Maine schools*. Maine Department of Education.
Maine Department of Education. (2024b). *Climate Education Professional Development Grant Pilot Program*. Office of Innovation, Maine Department of Education.
Mbah, M. F., Shingruf, A., & Molthan-Hill, P. (2022). Policies and practices of climate change education in South Asia: Towards a support framework for an impactful climate change adaptation. *NPJ Climate Action, 1*, 28. https://doi.org/10.1038/s44168-022-00029-0
McKenzie, M., & Benavot, A. (2024). *Climate change and sustainability in science and social science secondary school curricula*. UNESCO.
Meadows, D. H. (2008). *Thinking in systems: A primer*. (D. Wright, Ed.). Chelsea Green Publishing.
MECCE Project. (2020). *Announcing the launch of the Monitoring and Evaluating Climate Communication and Education* (MECCE) Project.
Meckler, L. (2023, September 6). Heat wave leads some schools to close or dismiss classes early. *The Washington Post*.
Meyer, J. W. (1977). The effects of education as an institution. *American Journal of Sociology, 83*(1), 55–77.
Meyer, J. W., Ramirez, F. O., & Soysal, Y. N. (1992). World expansion of mass education, 1870–1980. *Sociology of Education, 65*(2), 128–149.
Meyer, J. W., Bromley, P., & Ramirez, F. O. (2010). Human rights in social science textbooks: Cross-national analyses, 1970–2008. *Sociology of Education, 83*(2), 111–134.
Milman, O. (2023, August 10). Videos denying climate science approved by Florida as state curriculum. *The Guardian*. https://www.theguardian.com/us-news/2023/aug/10/florida-ron-desantis-climate-vidoes-school-curriculum
Mitchell, M. (2009). *Complexity: A guided tour*. Oxford University Press.
Monroe, M. C., Plate, R. R., Oxarart, A., Bowers, A., & Chaves, W. A. (2019). Identifying effective climate change education strategies: A systematic review of the research. *Environmental Education Research, 25*(6), 791–812.
National Council of Educational Research and Training. (2005). *National curriculum framework 2005*. NCERT.

National Environment Management Authority. (2018). National environmental education and communication strategy 2018–2022.
National Research Council. (1998). Children of the urban environment: The ecology of early development. In *Committee on Environmental Health Risks to Children, Environmental health risks to children* (pp. 65–90). National Academy Press.
Organisation for Economic Co-operation and Development. (2010). *Strong performers and successful reformers in education: Lessons from PISA for the United States.* OECD Publishing.
Organisation for Economic Co-operation and Development. (2013). *Synergies for better learning: An international perspective on evaluation and assessment.* OECD Publishing.
Organisation for Economic Co-operation and Development. (2025). *Trends shaping education 2025.* OECD Publishing.
Oppenheimer, M., & Petsonk, A. (2005). Article 2 of the UNFCCC: Historical origins, recent interpretations. In D. Freestone & C. Streck (Eds.), *Legal aspects of implementing the Kyoto Protocol mechanisms: Making Kyoto work* (pp. 19–41). Oxford University Press.
Paltsev, S., Schlosser, C. A., et al. (2025). *2025 Global Change Outlook.* MIT Center for Sustainability Science and Strategy. (https://cs3.mit.edu/publications/signature/2025-global-change-outlook)
Park, R. J., Behrer, A. P., & Goodman, J. (2021). Learning is inhibited by heat exposure, both internationally and within the United States. *Nature Human Behaviour, 5*(1), 19–27.
Park, R. J., Goodman, J., Hurwitz, M., & Smith, J. (2020). Heat and learning. *American Economic Journal: Economic Policy, 12*(2), 306–339.
Reimers, F. M. (2020a). *Educating students to improve the world.* Springer.
Reimers, F. M. (Ed.). (2020b). *Audacious education purposes: How governments transform the goals of education systems.* Springer.
Reimers, F. M. (Ed.). (2021). *Education and climate change: The role of universities.* Springer. https://doi.org/10.1007/978-3-030-57927-2
Reimers, F. M. (Ed.). (2024). *Schools and society during the COVID-19 pandemic: How education systems changed and the road ahead.* Springer Nature Switzerland.
Risbey, J. S., et al. (2008). The new climate discourse: Climate change as a challenge for risk management. In H. G. Brauch (Ed.), *Globalization and environmental challenges* (pp. 455–468). Springer.
Roston, E. (2023, September 9). As school started in the U.S., so did the school closures for heat. *Time.*
Rousell, D., & Cutter-Mackenzie-Knowles, A. (2020). A systematic review of climate change education: Giving children and young people a "voice" and a "hand" in redressing climate change. *Children's Geographies, 18*(2), 191–208.
Sabarwal, S., Venegas Marin, S., Spivack, M., & Ambasz, D. (2024). *Choosing our future: Education for climate action.* World Bank.
Sabbatini, C., & Beech, J. (2025). *The whole-school approach to climate change education in action: A year with Primary School No. 12 Mar Chiquita.* MECCE Project Funded Case Study.
Shepardson, D. P., Choi, S., Niyogi, D., & Charusombat, U. (2011a). Seventh grade students' mental models of the greenhouse effect. *Environmental Education Research, 17*(1), 1–17. https://doi.org/10.1080/13504620903564549
Shepardson, D. P., Niyogi, D., Choi, S., & Charusombat, U. (2009). Seventh grade students' conceptions of global warming and climate change. *Environmental Education Research, 15*(5), 549–570. https://doi.org/10.1080/13504620903114596
Shepardson, D. P., Niyogi, D., Choi, S., & Charusombat, U. (2011b). Students' conceptions about the greenhouse effect, global warming, and climate change. *Climatic Change, 104*(3–4), 481–507. https://doi.org/10.1007/s10584-009-9786-9
Shepardson, D. P., Niyogi, D., Roychoudhury, A., & Hirsch, A. (2012). Conceptualizing climate change in the context of a climate system: Implications for climate and environmental education. *Environmental Education Research, 18*(3), 323–352. https://doi.org/10.1080/13504622.2011.622839

Shapiro Ledley, T., Rooney Varga, J., & Niepold, F. (2017). *Addressing climate change through education.* Oxford University Press. https://doi.org/10.1093/acrefore/9780199389414.013.56

Singh, V. (2023). *Teaching climate change: Science, stories, justice.* Routledge.

State Council of China. (2011). *Outline of China's national plan for medium and long term education reform and development (2010–2020).* State Council of the People's Republic of China.

Sweden – Climate Change Communication and Education Country Profile. (2023). *In Climate change communication and education country profiles: Approaches to greening education around the world.* UNESCO Global Education Monitoring Report/MECCE Project.

Swedish National Agency for Education. (2011). *Curriculum for the compulsory school, preschool class and school-age educare 2011* (rev. 2022). Skolverket.

Tyack, D., & Cuban, L. (1995). *Tinkering toward utopia: A century of public school reform.* Harvard University Press.

UNESCO. (1975). The Belgrade Charter: A global framework for environmental education. UNESCO–UNEP.

UNESCO. (1977). *Intergovernmental Conference on Environmental Education, Tbilisi (USSR), 14–26 October 1977: Final report (ED/MD/49).* UNESCO.

UNESCO. (2005). *United Nations Decade of Education for Sustainable Development 2005–2014: The DESD at a glance.* UNESCO.

UNESCO. (2010a). *Education for Sustainable Development lens: A policy and practice review tool.* UNESCO.

UNESCO. (2010b). *Education for sustainable development: Achievements and future directions.* UNESCO.

UNESCO. (2014a). *Shaping the future we want: UN Decade of Education for Sustainable Development (2005–2014) final report.* UNESCO.

UNESCO. (2014b). *UNESCO roadmap for implementing the Global Action Programme on Education for Sustainable Development.* UNESCO.

UNESCO. (2017a). *Education for Sustainable Development Goals: Learning objectives.* UNESCO.

UNESCO. (2017b). *Changing minds, not the climate: The role of education.* UNESCO.

UNESCO. (2018). *Learning to live together sustainably: Curriculum perspectives.* UNESCO International Bureau of Education.

UNESCO. (2019). *Framework for the implementation of Education for Sustainable Development (ESD) beyond 2019 (ESD for 2030).* UNESCO.

UNESCO. (2020). *Education for Sustainable Development: A roadmap (ESD for 2030).* UNESCO.

UNESCO. (2021a). *Reimagining our futures together: A new social contract for education.* UNESCO.

UNESCO. (2021b). *Berlin Declaration on Education for Sustainable Development: Learn for our planet. Act for sustainability.* UNESCO.

UNESCO. (2021c). *Getting every school climate ready: How countries are integrating climate change issues in education.* UNESCO.

UNESCO. (2021d). *Learn for our planet: A global review of how environmental issues are integrated in education.* UNESCO.

UNESCO. (2022). *Youth demands for quality climate change education.* UNESCO.

UNESCO. (2023a). *Learning at risk: The impact of climate displacement on the right to education* [Global report]. UNESCO.

UNESCO. (2023b). *Climate change communication and education country profiles: Approaches to greening education around the world.* UNESCO.

UNESCO. (2023c). *Empowering learners and teachers for climate action.* UNESCO.

UNESCO. (2023d). *Climate change education: Education is crucial to promote climate action.* UNESCO.

UNESCO. (2023e). *UNESCO at COP28: Making education the long-term solution to the climate crisis.* UNESCO.

UNESCO. (2024a). *Greening every curriculum: Guidance to teach and learn for climate action.* UNESCO.

UNESCO. (2024b). *Green school quality standard: Greening every learning environment.* UNESCO.
UNESCO. (2024c). *Greening Education Partnership: Getting every learner climate ready.* UNESCO.
UNESCO. (n.d.). *Climate change education.* UNESCO. https://www.unesco.org/en/query-list/c/climate-change-education
UNESCO. (n.d.). *Man and the Biosphere Programme (MAB).* UNESCO. https://www.unesco.org/en/mab
UNESCO; United Nations University Institute for the Advanced Study of Sustainability; & UNESCO Bangkok. (2023). *Asia-Pacific regional synthesis: Climate change, displacement and the right to education.* UNESCO; UNESCO Bangkok; UNU IAS.
UNESCO Global Education Monitoring Report Team, & MECCE Project. (2023a). *Italy: Climate change communication and education country profile.* UNESCO.
UNESCO Global Education Monitoring Report Team, & MECCE Project. (2023b). *Scotland: Climate change communication and education country profile.* UNESCO.
UNESCO Global Education Monitoring Report Team, & MECCE Project. (2023c). *New Zealand: Climate change communication and education country profile.* UNESCO.
UNESCO Global Education Monitoring Report Team, & MECCE Project. (2023d). *Dominican Republic: Climate change communication and education country profile.* UNESCO.
UNESCO Global Education Monitoring Report Team, & MECCE Project. (2023e). *United States: Climate change communication and education country profile.* UNESCO.
UNESCO Global Education Monitoring Report Team, & MECCE Project. (2023f). *Mexico: Climate change communication and education country profile.* UNESCO.
UNESCO Global Education Monitoring Report Team, & MECCE Project. (2024). *Education and climate change: Learning to act for people and planet.* UNESCO.
UNESCO International Bureau of Education. (2016). *Training tools for curriculum development: A resource pack.* UNESCO IBE.
UNESCO International Bureau of Education. (2018). *Learning to live together sustainably: Curriculum perspectives.* UNESCO IBE.
UNESCO International Bureau of Education. (2022). *Transforming curriculum and learning for climate action.* UNESCO IBE.
UNESCO International Bureau of Education. (2023a). *UNESCO IBE: Transforming curriculum and greening education.* UNESCO IBE.
UNESCO International Bureau of Education. (2023b). *Greening education systems: A curriculum centred approach.* UNESCO IBE.
UNESCO International Bureau of Education. (2024a). *Greening the curriculum: Flagship initiative.* UNESCO IBE.
UNESCO International Bureau of Education. (2024b). *Greening schools and learning environments: A whole-school approach.* UNESCO IBE.
UNESCO–UNEVOC. (2021). *Skills development and climate change action plans: Enhancing TVET's contribution.* UNESCO–UNEVOC.
UNICEF. (2025). *Learning interrupted: Global snapshot of climate-related school disruptions in 2024.* UNICEF. https://www.unicef.org/reports/learning-interrupted-global-snapshot-2024
United Nations. (1948). *Universal Declaration of Human Rights.* United Nations.
United Nations. (1992). *Report of the United Nations Conference on Environment and Development (Rio de Janeiro).* United Nations.
United Nations. (2015). *Transforming our world: The 2030 Agenda for Sustainable Development.* United Nations.
United Nations. (2020). Education is key to addressing climate change. In *Climate solutions: Education* [Policy brief]. United Nations.
United Nations Climate Security Mechanism. (n.d.). *The climate dictionary.* United Nations.
United Nations Conference on Environment and Development. (1992). *Agenda 21: Programme of action for sustainable development.* United Nations.

United Nations Framework Convention on Climate Change. (1992). *United Nations Framework Convention on Climate Change*. UNFCCC.
United Nations Framework Convention on Climate Change. (2012). *Action for Climate Empowerment (ACE) work programmes and related decisions*. UNFCCC.
United Nations Framework Convention on Climate Change. (2015). *Paris Agreement*. UNFCCC.
United Nations Office for Disaster Risk Reduction. (2019). *Critical infrastructure including schools, health facilities and roads threatened by climate crisis*. UNDRR.
Venegas Marin, S., Schwarz, L., & Sabarwal, S. (2024). *The impact of climate change on education and what to do about it*. World Bank.
Waldrop, M. M. (1992). *Complexity: The emerging science at the edge of order and chaos*. Simon & Schuster.
World Bank. (2018). *World development report 2018: Learning to realize education's promise*. World Bank.
World Meteorological Organization. (2014). *Weather reports from 2050* [Video series]. WMO.
World Meteorological Organization. (2022). *It's warmer than average. But what is average?* WMO.
World Meteorological Organization, & Climate Central. (2017). *Summer in the cities* [Video series]. WMO & Climate Central.
Worth, K. (2021). *Miseducation: How climate change is taught in America*. Columbia Global Reports.
Xu, C., & Iyengar, R. (2023). *Climate change education: An Earth Institute sustainability primer*. Columbia University Press.

Open Access This chapter is licensed under the terms of the Creative Commons Attribution-NonCommercial-NoDerivatives 4.0 International License (http://creativecommons.org/licenses/by-nc-nd/4.0/), which permits any noncommercial use, sharing, distribution and reproduction in any medium or format, as long as you give appropriate credit to the original author(s) and the source, provide a link to the Creative Commons license and indicate if you modified the licensed material. You do not have permission under this license to share adapted material derived from this chapter or parts of it.

The images or other third party material in this chapter are included in the chapter's Creative Commons license, unless indicated otherwise in a credit line to the material. If material is not included in the chapter's Creative Commons license and your intended use is not permitted by statutory regulation or exceeds the permitted use, you will need to obtain permission directly from the copyright holder.

Chapter 2
Mapping the Terrain. Perspectives on Education and Climate Change

As the previous chapter discussed, the relationship between education and climate change is a two-way street. On the one hand, climate change influences opportunity to learn, through the first order effects of climate change on education discussed in the previous chapter. On the other hand, the second order effects include education curricula and programs that aim to help students develop skills to adapt to and mitigate the impact of climate change. In this case, the intended direction of the relationship goes from education to climate change and its expected impact. There are various perspectives about how students should be educated to adapt to and mitigate the impact of climate change. This chapter examines what is known about this two-way relationship between education and climate change and expands on these multiple perspectives on education and climate change.

2.1 First-order Effects: How Climate Change Impacts Education, and Approaches to Mitigation and Adaptation

Climate change affects education systems through what we term first-order effects: those pathways by which climatic hazards and slow-onset changes directly disrupt students, teachers and institutions. These include the immediate health and cognitive effects of heat or pollution on learning; the destruction or damage of school infrastructure; the interruption of instructional time through closures and displacement; and the cascading impacts of climate-induced poverty, illness and psychological distress on educational trajectories.

Globally, the scale of disruption is already substantial. In 2024 alone, an estimated 242 million children from pre-primary through upper secondary education experienced schooling disruptions linked to climate events (UNICEF, 2024). In some highly exposed systems the effects are routine and cumulative: in the Philippines, for

© The Author(s) 2026

F. Reimers and M. Wang-Aghania, *Educating for a Climate Changed Future*,
https://doi.org/10.1007/978-3-032-23256-4_2

example, public schools lose between 30 and 50 instructional days each year due to typhoons, floods, landslides and extreme heat (EDCOM II et al., 2024; Alabi et al., 2026; World Bank Group & Asian Development Bank, 2021).

These first-order impacts are not evenly distributed. They interact with pre-existing inequalities and institutional capacities, such that children in rural, coastal and informal urban settlements, especially those in low- and middle-income countries, bear a disproportionate share of lost learning time and foregone attainment (World Bank Group & Asian Development Bank, 2021; Leal Filho et al., 2023; Sprague et al., 2025). The World Health Organization's COP26 Special Report frames climate change as "the single biggest health threat facing humanity," emphasizing that climate-sensitive health risks (heat stress, extreme events, degraded air and water quality, food insecurity, vector-borne disease, and mental health problems) fall most heavily on disadvantaged populations and threaten to reverse progress in development and poverty reduction (World Health Organization, 2021). Because educational opportunity is tightly coupled to health, income and social protection, climate change functions as a "threat multiplier" for existing educational disparities.

For analytic clarity, we can distinguish three intertwined pathways: (1) direct impacts on learning and educational outcomes; (2) cumulative and life-course effects mediated through livelihoods, poverty and health; and (3) infrastructure and systemic disruption. Across all three, mental health and learning continuity are central.

2.1.1 *Direct Impacts on Learning*

A first body of evidence documents how climate-related exposures, particularly heat and air pollution, impair cognition, concentration and test performance. Roach and Whitney (2022), analyzing data for elementary and middle-school students in the United States, find that increasing average temperatures and days of extreme heat reduce achievement in mathematics and English/Language Arts; each additional day above 100°F is associated with measurable declines in standardized test scores, with especially large effects in historically cooler areas. International analyses similarly show that learning is highly sensitive to thermal stress when classrooms lack adequate cooling, ventilation or shading.

Sprague et al. (2025), in a weight-of-evidence scoping review of 31 studies on childhood climate exposures and educational outcomes, report that all 31 studies find negative impacts on at least one educational dimension. These include lower achievement in humanities, mathematics and science; impairments in executive functioning; reduced attention; higher absenteeism; and delays in key advancement milestones such as grade promotion and completion. Climate change thereby erodes learning not only through spectacular disasters but also through routine, sub-optimal conditions in which teaching and learning occur.

Evidence from large-scale assessments reinforces this picture. Park et al. (2021), examining PISA performance in the United States, find that cumulative exposure to hot school days is associated with lower scores, with the largest effects for students in lower-income districts, younger students, and racial and ethnic minorities (cited in Prentice et al., 2024). Prentice et al. (2024) also review studies showing that exposure to wildfire smoke in Indonesia reduces completed years of schooling, and that short-term wildfire-related school closures in California depress academic performance.

In highly climate-exposed systems, repeated interruptions compound these direct effects. In the Philippines, schools on average lose several weeks of face-to-face instruction each year to typhoons, flooding and heat waves (EDCOM II et al., 2024; World Bank Group & Asian Development Bank, 2021). Herrera-Almanza and Cas (2020) and Cas (2016) show that typhoon-related disruptions, particularly in schools with weaker infrastructure and fewer instructional resources, are associated with long-term losses in human capital. Even where "alternative learning modalities" such as printed modules, television lessons, radio programs or online platforms are activated during closures, implementation gaps, limited connectivity, device shortages, and weak guidance for teachers and students often mean that these days do not translate into equivalent learning (Lagmay & Rodrigo, 2022). In Lagmay and Rodrigo's (2022) study of online learning participation during extreme weather, for instance, student engagement fell sharply on days with severe rainfall, power interruptions or connectivity problems, underscoring how climate conditions undermine the effectiveness of remote instruction.

Mental health is a further direct pathway. Burke et al. (2018) review evidence on the psychological effects of climate change on children and conclude that both acute disasters (e.g., typhoons, floods, wildfires) and chronic stressors (e.g., drought, heat) increase risks of post-traumatic stress disorder, depression, anxiety and sleep disorders. These conditions impair emotion regulation, cognition, attention and behavior, with clear implications for schooling and learning. Children in developing countries are particularly vulnerable because climate shocks intersect poverty, weak health and mental-health services, and limited social protection.

Finally, climate change can shape students' emotional relationship to schooling and to the future. Jones and Davison (2021) find that educational experiences about climate change sometimes leave students feeling overwhelmed or disempowered, which may dampen long-term engagement with the issue if not addressed through pedagogies that promote agency and collective efficacy. These emotional dynamics are part of the first-order relationship between climate and education, even though they also blur into the second-order question of how climate is taught.

2.1.2 Cumulative and Life-course Effects

First-order effects are not confined to short-term disruptions. They accumulate over the life course as climate change reshapes livelihoods, health and household decisions about education. Extreme events and slow-onset changes disproportionately affect climate-sensitive sectors such as rain-fed agriculture, fisheries and tourism. When yields fall, livestock die, fisheries collapse or tourism declines, household incomes contract. Families may respond by reducing educational expenditures, pulling children out of school, increasing their work responsibilities, or delaying transitions to higher levels of education.

Prentice et al. (2024) synthesize studies across multiple countries that trace these channels. In India, Garg and colleagues (cited in Prentice et al., 2024, p. 215) find that children's test scores decline as average temperatures rise, in part because heat reduces yields of heat-sensitive crops, increases poverty and malnutrition, and thereby undermines learning. A study of a typhoon in the Philippines shows that the event reduced family wealth and educational investment, leading to lower academic achievement, higher grade repetition and lower completion rates, especially for girls, poorer children, older children and those with less familial support (Deuchert & Felfe, 2015, cited in Prentice et al., 2024, p. 217). Herrera-Almanza and Cas (2020) further document that typhoon exposure during school age has long-run negative effects on educational attainment and earnings, and that these losses are larger where school infrastructure and instructional resources are weaker.

Floods have been found to depress attendance, enrollment and achievement in contexts as varied as Thailand, Ethiopia, India, Peru, Vietnam and Pakistan (Prentice et al., 2024, p. 217). In Zimbabwe, flood-related damage to school sanitation facilities increased illness and absenteeism (ibid.). Droughts increase the incidence of diseases such as typhoid, cholera and diarrhea and reduce agricultural incomes; studies in Nigeria, Iran, Kenya and Mexico link these conditions to lower school attendance and progression (Prentice et al., 2024, p. 217). Leal Filho et al. (2023) highlight that climate-related shocks to income and health are especially damaging for children's educational attainment in low-income and developing countries, where social protection and public services are thin.

These negative effects often interact cumulatively over time. Early-life exposure to malnutrition, disease, trauma or repeated school disruptions can sensitize children to later shocks, making subsequent hazards more damaging for learning and socio-emotional development (Prentice et al., 2024, p. 214; Burke et al., 2018). In this sense, climate change alters the developmental ecology within which educational trajectories unfold. Understanding first-order effects requires a life-course perspective that traces how multiple exposures and vulnerabilities intersect.

Taken together, this literature shows that first-order effects are not episodic shocks but cumulative forces that reshape educational trajectories over time, particularly for children already facing structural disadvantage.

2.1.3 *Infrastructure, System Disruption and Learning Continuity*

A third set of pathways operates through damage to school infrastructure, disruption of operations, and weaknesses in system-level continuity planning. Natural disasters, typhoons, cyclones, hurricanes, floods, landslides, wildfires and earthquakes destroy or damage school buildings, furniture, laboratories and libraries, as well as basic services such as water, sanitation and electricity. A study for the UN Office for Disaster Risk Reduction found that in 83 countries an average of 3200 schools per year were damaged or destroyed by weather-related events, compared to 412 health facilities (UNDRR, 2019). Damaged schools may be closed for weeks or months while repairs are completed; in many countries they are also used as evacuation centers, further delaying the resumption of classes and diverting limited resources from education to shelter (ASEAN, 2015; Ramos, 2024).

The Philippine case provides a vivid illustration of these dynamics. Many public school buildings are located in multi-hazard zones, coastal floodplains, steep slopes prone to landslides, or low-lying urban districts with poor drainage, and were not designed to withstand repeated typhoons, flooding and seismic events (Nassirpour et al., 2018; Pacific Disaster Center, 2021; EDCOM II et al., 2024). Roof systems, windows and foundations are often deteriorated, especially in buildings constructed before 2000 (Nassirpour et al., 2018). When storms hit, roofs are torn off, walls and floors are saturated, and access roads are blocked. Schools are frequently designated as emergency shelters, meaning that even buildings that remain structurally sound cannot immediately resume instructional use (Ramos, 2024). A scenario-based recovery analysis of education infrastructure in the Philippines suggests that such structural vulnerabilities are a major driver of prolonged closures and that risk-informed, resilient design can significantly reduce recovery times (Opabola & Galasso, 2024; Pal et al., 2023).

Infrastructure fragility intersects with digital and organizational capacities. As the COVID-19 pandemic showed, the ability to shift to remote or blended learning hinges on connectivity, devices, content and teacher preparation. In the Philippines, a review of the Learning Continuity Plan found that insufficient devices, poor data access and unreliable internet connections severely constrained the implementation of alternative learning modalities. School heads in rural Agusan del Sur described limited communication channels and resource shortages that impeded basic coordination during emergencies (Angwas, 2025). Where connectivity is weak, schools must rely on printed modules, radio programs or other low-bandwidth options (Lagmay & Rodrigo, 2022; You, 2024). Similar patterns are documented in peri-urban Dhaka (Kabir et al., 2021) and other parts of South and Southeast Asia (UNICEF East Asia and Pacific Regional Office, 2025; ADB, 2021).

Governance structures shape how these vulnerabilities translate into closures and learning loss. In many systems, responsibilities for disaster risk management and education are fragmented. In the Philippines, for example, the Department of Education (DepEd) manages curriculum and learning modalities through a centralized hierarchy, while local chief executives (mayors and governors) have legal authority to suspend classes under the 1991 Local Government Code (Iuchi et al., 2019). This split leads to unclear triggers for suspensions and for activating alternative modalities. Without standardized thresholds tied to hazard data or a national, tiered continuity framework, decisions are often made ad hoc, based on political or parental pressure, leading to inconsistent practices across regions and frequent "default" closures. Chua Reyes et al. (2021) describe how, under conditions of poverty, disasters and what they call "cultures of crisis," school leaders are left to improvise responses with limited guidance, contributing to highly uneven learning continuity.

These governance challenges are not unique to the Philippines. Comparative work in central Vietnam (Tong et al., 2012), Thailand (Ariyasirichot, 2021), and across Southeast Asia (Comfort, 2012; OECD, 2018) shows that continuity improves when hazard data, early-warning systems and school-level planning are aligned across agencies and when roles and decision rules are clearly defined. Where such alignment is absent, even relatively small-scale hazards can trigger disproportionate instructional losses because schools close preemptively or remain closed longer than necessary.

2.1.4 System-level Adaptation to First-order Effects: Learning Continuity Strategies

Because first-order effects operate through these structural vulnerabilities, they are amenable, at least in part, to deliberate adaptation. Education systems cannot prevent typhoons or heat waves, but they can reduce avoidable learning loss by strengthening learning continuity: the capacity to preserve predictable instructional time and to maintain meaningful learning through alternative modalities when face-to-face instruction is disrupted. Local school systems can also adapt to the effects of climate change modifying school calendars to recover instructional time lost because of climatic disruptions.

The Philippine policy analysis is instructive in this regard. Alabi et al. (2026) propose a set of complementary measures, sequenced over time, which illustrate broader principles:

Short-term, context-specific operational packages can reduce closures and accelerate reopening. In urban, low-lying divisions, an "urban heat and flood continuity package" combines modest calendar adjustments (e.g., shifting school breaks into peak heat months, altering daily schedules during heat waves) with low-cost ventilation, shading and drainage retrofits. Evidence from Vietnam, Indonesia, Thailand and the Philippines suggests that such measures can reduce heat-related lost instructional days by 30–50% within 1–2 years in targeted areas (Cooper et al., 2003; Akita & Miyata, 2021; Preña & Labayo, 2024; UNICEF East Asia and Pacific Regional Office, 2025). In rural and coastal divisions, a "rapid reopening and offline learning package" prioritizes repairs to roofs, windows, drainage and access routes, pre-positions materials and temporary structures, and expands low-bandwidth modalities (printed modules, radio, offline digital content). Scenario-based analyses and post-disaster studies indicate that such strategies can shorten closures by 10–15 days after severe events and reduce curriculum loss, especially when targeted to the most disruption-prone divisions (Akita & Miyata, 2021; Kabir et al., 2021; Opabola & Galasso, 2024; You, 2024).

Long-term investments in climate-resilient infrastructure address the structural drivers of repeated closure. Hazard-adaptive design and retrofit standards, elevated platforms and improved drainage in flood-prone areas, stronger structural frames and roof connections in typhoon-exposed zones, enhanced ventilation and shading in high-heat regions, can substantially reduce damage and downtime (ASEAN, 2015; Kabir et al., 2021; Pal et al., 2023). In the Philippines, the Typhoon-Resistant School Building Program has been associated with gains of 0.23–0.32 years of educational attainment in affected communities, suggesting that resilient construction can improve both safety and long-run learning (Cas, 2016; Herrera-Almanza & Cas, 2020; DepEd, 2024).

National tiered learning continuity frameworks can bring greater coherence and equity to instructional decisions under hazard conditions. Tong et al. (2012) and Alabi et al. (2026) describe frameworks that define a small number of hazard tiers (e.g., normal, localized disruption, division-wide disruption, system-wide emergency), each linked to pre-specified instructional responses and minimum learning expectations. When tied to meteorological and disaster-risk alerts, such frameworks reduce reliance on ad hoc judgement, clarify when to maintain face-to-face instruction with precautions and when to activate blended or low-connectivity modalities, and make practices more predictable for students, teachers and families (Comfort, 2012; Iuchi et al., 2019; Lagmay & Rodrigo, 2022; UNICEF, 2024; GPE, 2024).

Integration with regional early-warning and risk-analytics platforms offers additional lead time for anticipatory action. By linking education ministries to multi-hazard early-warning systems in ASEAN and ESCAP, for example, systems such as DepEd can identify which regions and school clusters are likely to be affected by upcoming typhoons, extreme rainfall or heat events and activate continuity plans earlier and more precisely (World Bank Group & Asian Development Bank, 2021; UNICEF East Asia and Pacific Regional Office, 2025; ADB, 2021).

The emerging evidence examined suggests that while climate change will continue to generate unavoidable shocks, education systems are not powerless. By investing in resilient infrastructure, clarifying governance arrangements, strengthening low-bandwidth learning options, and adopting coherent continuity frameworks, they can reduce the conversion of climatic hazards into lost learning and foregone futures, especially for those children who are currently most at risk.

2.2 Second-order Effects: Educating Students About and for Climate Change

If first-order effects describe how climate change disrupts education, second-order effects capture the converse: how education can, in turn, shape climate trajectories and societies' adaptive capacities. Here the causal arrow runs primarily from education to climate-relevant knowledge, behaviors, institutions and technologies.

Recent syntheses converge on three broad mechanisms through which education influences climate outcomes (Sabarwal et al., 2024; Anderson, 2012; Monroe et al., 2019):

- by changing what people know, value and do (pro-climate behaviors, political choices, social norms).
- by providing "green skills" and innovation capacity for low-carbon, climate-resilient economies.
- by building the adaptive capacities of individuals, communities and institutions, especially in highly vulnerable regions.

Integrative frameworks for climate change education span formal, non-formal and informal learning settings, emphasize systems thinking and systems outcomes, and highlight teachers and learners as central agents in climate response (Chang, 2014, 2023; Xu & Iyengar, 2023). Practitioner-oriented guides translate these ideas into NGSS-aligned classroom practice for secondary science teachers and argue for justice-centered, intersectional approaches to climate teaching anchored in the climate emergency (Lê, 2025).

At the same time, comparative reviews of education policy and curriculum remind us how under-realized this potential currently is. UNESCO's recent global review of 78 sector plans and curriculum frameworks from 46 countries finds that although 92% of documents mention environmental themes at least once, 45% devote little-to-no substantive attention to them; climate change appears in fewer than half of documents and biodiversity in only 19%, with by far the greatest emphasis on generic "environment" or "sustainability" (UNESCO, 2019, 2021). Teacher education lags even further behind: over one-third of surveyed teachers reported no exposure to environment-related content in either pre- or in-service training, and where environmental themes are present they are heavily concentrated in biology, general science and geography (UNESCO, 2021). Xu and Iyengar (2023) likewise emphasize that climate education opportunities are constrained by systemic barriers and inequalities, arguing that high-quality climate learning is unevenly distributed and that bringing marginalized voices, communities and knowledge systems into climate education is essential.

In some contexts, the challenge is not only absence but distortion. Worth's *Miseducation* documents how U.S. state standards, textbooks, school boards and fossil-fuel-linked actors have produced what she calls a red–blue divide in climate education, with students in some states receiving robust science-based instruction while others encounter "both-sides" narratives, denialist materials and textbook language that casts unwarranted doubt on the scientific consensus (Worth, 2021). This work underscores that climate change education is not simply a technical project of content insertion, but a politically contested terrain in which powerful interests seek either to dilute or to reshape what is taught.

Against this backdrop, the educational responsibilities that follow from the three mechanisms above are understood in diverse and sometimes conflicting ways. Climate change education (CCE) is a crowded and contested space. We distinguish at least five co-existing narratives through which the role of education in climate change is constructed and debated:

- Climate literacy: knowing is empowering.
- Education for climate action: from knowing to doing.
- Education for the Green Economy: skills for a sustainable future.
- Education for sustainable development: integrative and holistic approaches.
- Critical and decolonial climate education: justice, power and plurality.

These narratives are not mutually exclusive, but neither can they be distinctly synthesized. For example Education for climate action integrates climate literacy, and education for sustainable development incorporates skills for a sustainable future. Differences among these narratives are primarily of emphasis, each foregrounds particular priorities, theories of change and conceptions of agency. Recognizing this mosaic is helpful: it shifts attention from asking "what is the right model?" to asking "which combinations of goals and strategies are most appropriate in which contexts, and for whom?" A robust educational response to climate change will require pluralistic dialogue and critical reflection across these narratives rather than the triumph of any single one because robust coalitions are more likely to emerge from dialogue than from the divisiveness which excessive emphasis on the edges that separate those narratives might engender. The UNESCO Futures of Education report, for example, embodies elements of all five: it emphasizes scientific literacy and skills, systems thinking and holistic ESD, and justice-centered, decolonial perspectives (UNESCO, 2021).

What follows briefly characterizes each narrative, integrating available evidence on its potential, limits and current gaps, and situating it within recent global reviews of environmental education, green skills and CCE-for-ESD.

2.2.1 Climate Literacy: Knowing is Empowering

The climate-literacy narrative is perhaps the most familiar and institutionally embedded. It frames education's primary climate responsibility as epistemic: to cultivate a scientifically literate public that understands the basic workings of the Earth system, the evidence for anthropogenic climate change, and the range of possible impacts and responses. The underlying assumption is that if people understand science, they will be better able, and more inclined, to make informed personal, economic and political choices. To a great extent the assumption that understanding of science is essential is incorporated in all narratives, differences are about how much should such understanding encompass, and how should learning across disciplines and attention to local conditions be handled in learning climate science.

This narrative is illustrated in frameworks such as the US Global Change Research Program's Climate Literacy: The Essential Principles of Climate Science and initiatives like CLEAN, which specify core concepts (e.g., the greenhouse effect, the carbon cycle, feedbacks), practices (e.g., interpreting graphs, modeling, hypothesis testing) and learning progressions. Research shows that, across 96 countries and nearly one million students, educational attainment is the single strongest predictor of climate-change awareness: an additional year of schooling increases climate awareness by an estimated 8.6% (Sabarwal et al., 2024). People with more education are more likely to regard climate change as a serious threat and to understand its anthropogenic causes (Pew Research Center, 2019; Leiserowitz et al., 2013).

Chang's Climate Change Education: Knowing, Doing and Being (2014, 2023) offers a complementary synthesis from the perspective of geography and environmental education. The book defines CCE for the "global citizen," surveys how climate appears in formal curricula, analyzes what students know about climate change, and discusses pedagogies and assessment strategies to support more robust understanding. In doing so, it explicitly connects climate literacy to teacher readiness and to the design of climate-relevant learning progressions, arguing that fragmented and episodic treatment of climate topics is unlikely to yield deep understanding (Chang, 2014, 2023).

Global policy analysis, however, reveals that climate literacy is unevenly institutionalized. UNESCO's reviews of national curricula and education sector plans find that "environment" and generic "sustainability" receive far more attention than climate change or biodiversity; only 47% of policy and curriculum documents mention climate change at all, and when they do, coverage is typically cursory (UNESCO, 2019, 2021). Environmental themes are most often located in science, biology and geography; cross-curricular approaches remain more rhetoric than reality, and climate content is rare in early grades despite evidence that conceptions of weather, climate and human–environment relations begin forming well before secondary school (UNESCO MGIEP, 2019; UNESCO, 2021). Xu and Iyengar (2023) similarly argue that while climate change education is increasingly recognized as a priority, much of what currently exists is shallow and compartmentalized, and that systems thinking and attention to structural inequalities are needed if climate literacy is to be meaningful.

Mochizuki and Bryan argue that climate literacy is best understood within a broader framework of "Climate Change Education for Sustainable Development" (CCESD), which links scientific understanding of climate processes with knowledge of mitigation and adaptation strategies, local environmental conditions, disaster risk reduction and questions of responsibility, justice and rights (Mochizuki & Bryan, 2015). From this perspective, literacy involves not only "learning to know" climate science, but also "learning to do" (applying knowledge to mitigation and adaptation), "learning to live together" (grappling with interdependence, unequal vulnerabilities and ethical responsibility) and "learning to be" (developing the dispositions needed to cope with uncertainty, loss and change).

Empirical work also shows that students often hold robust but scientifically inaccurate mental models of climate change (see Chapter 6 for a detailed overview on student knowledge). In Singapore, for example, Chang and Pascua (2015) found that Grade 9 students' understandings of climate change comprised incomplete and incorrect elements organized into coherent mental models that students experienced as internally consistent. Common misconceptions mirrored international patterns:

confusing the greenhouse effect with ozone depletion, misunderstanding greenhouse gases and the nature of trapped heat, and linking climate change to unrelated environmental issues. Because these models are coherent, they can be resistant to instruction. Using an intervention-oriented evaluation design in a Singapore secondary school, Pascua and Chang (2015) showed that students' conceptual understanding improved substantially only after a geography teacher explicitly adopted direct, iterative refutation of misconceptions, informed by diagnostic data. This work illustrates that climate literacy initiatives must systematically diagnose and address misconceptions, not simply add more information.

Empirical and review work on curriculum further underscores that contemporary conceptions of climate literacy go beyond lists of facts. They emphasize integrated, interdisciplinary learning in which climate issues provide authentic contexts for STEM and social-science inquiry. Studies of integrated STEM education highlight three recurring elements: interdisciplinary connections, ambitious instructional practices, and engagement with real-world problems (Holmlund et al., 2018; Ring et al., 2017; Dare et al., 2021). When climate change is used as a unifying "real-world problem," students have opportunities to practice data analysis, systems thinking and design, and to connect scientific concepts to lived experience (Maspul, 2024).

Systematic reviews reinforce some key design principles. Monroe et al. (2019), analyzing 49 empirical studies of climate and environmental education interventions, identify strategies consistently associated with positive outcomes in knowledge, attitudes and, in some cases, self-reported behaviours:

- focusing on personally relevant and meaningful information rather than abstract global threats.
- using active, inquiry-based and collaborative pedagogies rather than purely didactic instruction.
- engaging learners in deliberative discussions, where they explain and defend their ideas and confront misconceptions.
- creating opportunities to interact with scientists and authentic data.
- explicitly surfacing and addressing misconceptions (for example, conflation of ozone depletion and the greenhouse effect).
- involving students in school or community projects that contextualize climate phenomena and potential solutions.

Anderson's (2012) review of climate change education similarly concludes that climate literacy is best fostered through sustained, active learning and cross-disciplinary curricula that connect scientific concepts to local, tangible problems and to individual behavior.

Yet both educational-quality and CCE reviews converge on persistent limitations. First, knowledge alone does not automatically translate into behavior or political engagement. Information-heavy interventions that lack attention to motivation, self-efficacy, social norms and structural barriers often produce modest or short-lived changes (Anderson, 2012; Monroe et al., 2019). Second, as UNESCO's policy analyses emphasize, most systems still treat environmental learning primarily as a cognitive endeavor: references to ESD-related themes focus overwhelmingly on knowledge; socio-emotional and behavioral dimensions are much less prominent (UNESCO, 2019, 2021). Chang (2014, 2023) and Xu and Iyengar (2023) both argue that this imbalance is a central weakness of current climate change education and call for approaches that integrate understanding with opportunities to act and to reflect on values and identities.

Third, foundational learning deficits sharply constrain climate literacy. Globally, an estimated 70% of ten-year-olds in low- and middle-income countries cannot read and understand a simple text (World Bank et al., 2022). Without literacy and numeracy, more advanced climate concepts and data-rich inquiry are inaccessible. For this reason, Sabarwal et al. (2024) and the ILO's green-skills work both argue that strengthening foundational skills is itself a powerful climate strategy: it underpins climate literacy, green skills and individuals' adaptive and economic capacities (ILO, 2019).

In short, climate literacy remains a necessary but not sufficient condition for meaningful participation in climate challenges. It provides essential cognitive infrastructure for other, more action-oriented or justice-focused narratives. CCESD perspectives suggest that climate literacy efforts are most promising when explicitly connected to broader questions of sustainable development, justice and futures, and when they cultivate not only understanding but also the capacities to cope, care and act (Chang, 2014, 2023; Mochizuki & Bryan, 2015; Xu & Iyengar, 2023).

2.2.2 *Education for Climate Action: From Knowing to Doing*

The education-for-climate-action narrative extends climate literacy from "knowing" to "doing," and, ultimately, to "becoming." It views education as an active lever for mitigation and adaptation, not merely as a transmitter of climate facts. The underlying theory of change is behavioral and civic: carefully designed learning experiences can shift everyday practices, social norms and political choices in ways that cumulatively reshape emissions and resilience pathways.

Drawing implicitly on what Kwauk and Casey describe as "green life skills," generic, cross-cutting competences such as critical thinking, collaboration, coping with uncertainty and empathy, this narrative emphasizes the importance of equipping learners not only with climate knowledge but also with the skills and dispositions needed to apply that knowledge in complex, contested real-world contexts (Kwauk & Casey, 2021).

Chang (2014, 2023) foregrounds this dimension in his "doing" and "being" components of CCE, arguing that climate change education should help learners not only understand climate science but also imagine and enact individual and collective responses. Xu and Iyengar (2023) likewise advocate for action-oriented, justice-centered climate teaching grounded in systems thinking, and show how such approaches can be implemented across formal, non-formal and community settings. Lê's (2025) guide for grades 6–12 provides a concrete instantiation of this narrative in the U.S. context: working within the NGSS framework, she encourages secondary science teachers to use climate change as an "anchor phenomenon" for unit design, to address environmental and climate justice explicitly, and to support students as agents of change in their own communities.

Empirical evidence supports key elements of this narrative. Sabarwal et al. (2024) synthesize cross-country analyses showing that:

- Additional years of schooling are associated with higher self-reported pro-climate behaviors (e.g., energy conservation, sustainable consumption) and greater willingness to change lifestyles in response to climate change.
- Education is particularly important for adaptation behaviors in low- and middle-income countries: more educated farmers, for example, are more likely to adopt drought-resistant crops, adjust planting calendars, diversify livelihoods and purchase insurance (Aryal et al., 2021).
- In Ethiopia, completing six years of schooling increases the likelihood that a farmer will adopt climate-adaptive practices by around 20%; similar patterns are documented across ten African countries.
- More educated citizens are more likely to support and vote for environmental policies and green parties; in Europe, an additional year of education increases green voting by an estimated 3.6% points.

Education can also catalyze collective action. Studies in the US and elsewhere show that higher education is associated with greater civic and political engagement, including voting, protest participation and policy advocacy (Sabarwal et al., 2024). Climate-focused curricula that weave in media literacy and argumentation can strengthen students' capacity to evaluate claims, counter misinformation and communicate science-based positions (Lambert & Bleicher, 2014; Rooney-Varga et al., 2014).

Importantly, education for climate action can have immediate effects, not only long-term ones. Interventions with children and adolescents have been shown to influence their parents' beliefs and behaviors, creating intergenerational "spillovers." In India, a school-based environmental program increased not only students' pro-environmental behaviors but also those of their parents by nearly 13% (Sabarwal et al., 2024). In the United States, climate education for middle-school students increased parents' concern about climate change, with the strongest effects among initially skeptical conservative parents (Stevenson et al., 2014, Lawson et al., 2019). Similar processes have been documented around recycling behaviors in the UK and disaster preparedness in Indonesia and Thailand (Kabir et al., 2021; Taber & Taylor, 2009).

Monroe et al.'s (2019) review identifies a cluster of strategies that are particularly effective when the goal is to change behavior, not only attitudes:

- personally relevant, actionable content that links climate issues to everyday decisions (transport, energy, food).
- locally focused problem-solving and citizen-science projects that make abstract phenomena tangible.

- explicit cultivation of self-efficacy and agency (showing that individual and collective actions matter).
- opportunities for deliberation and social learning about trade-offs and values.
- engagement in school or community projects that allow learners to design, implement and observe concrete mitigation or adaptation measures.

Anderson (2012) similarly emphasizes motivational framing: messages that highlight feasible solutions and align with learners' values, rather than emphasizing catastrophe, tend to foster greater engagement and intent to act, and to reduce paralysis and eco-anxiety.

However, global evidence suggests that education for climate action is far from mainstream. UNESCO's reviews of curricula and the ILO's analysis of NDCs both show that, although most countries now mention education or public awareness in their climate commitments, very few specify concrete, resourced strategies for building climate-action competences in schools or for adults (ILO, 2019; UNESCO, 2019). The Learn for our planet study finds that when environmental themes appear in policy and curriculum documents they are overwhelmingly framed in cognitive terms; references to socio-emotional learning and action competences remain scarce (UNESCO, 2021). Interviews and surveys reveal similar patterns: teachers and stakeholders recognize the importance of student participation in environmental activities, yet curricular guidance and assessment rarely prioritize these dimensions (UNESCO, 2021).

Moreover, education systems are only weakly mobilized for climate action. Less than one-third of NDCs mention climate education, and less than one-quarter mention green skills; only around 1.5% of climate finance currently goes to education (Sabarwal et al., 2024). Consistent with UNESCO's findings that over one-third of teachers receive no environment-related training (UNESCO, 2021), Sabarwal et al. (2024) report that across six low- and middle-income countries, 87% of teachers say they include climate topics in their lessons, but 71% answer at least one basic climate question incorrectly. Chang (2014, 2023) similarly emphasizes that teachers are key stakeholders in climate change discourse and that their readiness and professional development are critical determinants of whether CCE moves from policy to practice.

These gaps echo broader critiques that climate-action education often underestimates the adaptive and transformative dimensions of the climate challenge. Kwauk and Casey distinguish between:

- skills for green jobs (specific technical skills that support low-carbon sectors).
- green life skills (generic skills, problem-solving, collaboration, coping with uncertainty, which enable greener decisions and behaviors across jobs and life domains).
- skills for green transformation (capacities for critical analysis of power, coalition-building, political agency and systems change).

Much of the education-for-climate-action narrative focuses on the first two, with limited attention to the third. Yet climate change is as much an adaptive challenge, requiring changes in worldviews, social norms and institutions, as a technical one

(Kwauk & Casey, 2021). Without explicit work on political literacy, justice and structural change, education risks exhorting individuals to "do their bit" while leaving underlying drivers of emissions and vulnerability untouched.

In light of this, Sabarwal et al. (2024) propose three near-term priorities for harnessing schools for climate action that are consistent with both CCE-ESD and green-skills framings:

- Foundations first: secure literacy and numeracy, and then use climate topics to teach foundational skills, rather than "bolting on" climate content that crowds them out.
- Practical, actionable and contextual curriculum: integrate climate issues into existing subjects in locally meaningful ways, and emphasize behaviors and solutions, not only causes and impacts.
- Teacher capacity: strengthen teachers' own climate knowledge, pedagogical strategies and socio-emotional skills, and provide high-quality resources.

The climate-action narrative underscores that education can shape not only what people know about climate change, but how they think, feel, collaborate and act in the face of it. The challenge is to design interventions that move beyond awareness to agency, without over-burdening schools and teachers, and while linking everyday practices to larger questions of justice and structural transformation (Chang, 2014, 2023; Lê, 2025; Xu & Iyengar, 2023).

2.2.3 *Education for the Green Economy: Skills for a Sustainable Future*

A third narrative situates education's climate mission within the broader transition to a green and just economy. Here the focus is on "skills for the green transition": the technical, cognitive and socio-emotional competences needed to decarbonize production and consumption, adapt infrastructure and services, and ensure that structural changes are fair.

Global analyses estimate that green transitions will require skilled workers for around 100 million new jobs, substantial up-skilling for most existing jobs, and reskilling for some 78 million jobs likely to disappear (ILO, 2019; Sabarwal et al., 2024). Yet skill shortages already constrain transitions: in India, 60% of surveyed energy-sector employers report difficulties finding workers with adaptation and mitigation skills; in Indonesia, 76% of manufacturing firms see changes in existing jobs as the biggest adjustment needed for greening their business (Sabarwal et al., 2024).

This implies that education for the green economy must extend well beyond specialized technical training for new "green jobs." ILO's Skills for Greener Future and the earlier discussed new green learning agenda from Kwauk & Casey (2021) converges on a broader understanding of green skills that includes transversal, civic and adaptive capacities alongside occupation-specific competencies. In this sense, green economy agendas are moving closer to education-for-action frameworks: both emphasize systems thinking, agency, collaboration and the ability to navigate uncertainty and contestation. The shift reflects growing recognition that decarbonization and adaptation are not only technical challenges, but social ones, requiring education systems to prepare learners not just for emerging occupations, but for participation in shaping just transition pathways.

Contrary to common myths, green skills are not confined to high-income countries, advanced degrees, STEM occupations or "green" sectors. Using job-posting and labor-force data from Brazil, Egypt, India, Kenya and the Philippines, Sabarwal et al. (2024) show that:

- Demand for green skills is already substantial in low- and middle-income countries; for example, in Kenya, about 8% of online job postings between 2022 and 2023 required at least one green skill.
- Many green jobs do not require tertiary or STEM qualifications; in Egypt and Kenya, 10–20% of green job postings require only lower-secondary or primary education.
- Green skills are demanded across a wide range of sectors, including food and beverage services, the arts and entertainment, retail, education and finance, not only energy, construction or transport.
- Many green skills are cross-sectoral and transferable (e.g., sustainable procurement, waste management, environmental auditing, logistics efficiency).

Xu and Iyengar (2023) add that climate change education must also prepare "future workforces" in a broader sense, emphasizing that climate literacy, systems thinking and awareness of social and environmental impacts will shape a wide array of careers, not just those explicitly labeled "green." Their primer highlights systemic barriers and inequalities in access to climate learning and stresses that diversifying who participates in climate-relevant education is itself a prerequisite for a just transition.

The Skills for a Greener Future study further highlights that green job opportunities are "closer and bigger than many think," but also unpredictable and potentially inequitable: demand for specific skills evolves rapidly and differently across countries; women and marginalized groups are under-represented in green jobs, especially in leadership roles (ILO, 2019). Kwauk and Casey (2021) emphasize that unless green-economy and green-skills agendas are explicitly gender-transformative, addressing norms, opportunity structures and structural discrimination, they risk reproducing existing inequalities in a slightly greener labor market.

To respond, Sabarwal et al. (2024) and the ILO propose two time-horizons:

- In the short run, expand information and access by providing students and workers with clear labor-market information about green skills and returns; offering flexible, stackable short courses and apprenticeships; and partnering with employers to co-design training.
- In the medium run, foster adaptable students and systems by ensuring strong foundational learning; strengthening transferable cognitive and socio-emotional skills; building flexible pathways between general and vocational education; and

improving alignment among ministries, training providers and employers (ILO, 2019).

However, policy analysis suggests that coordination between environmental, economic and skills policies is weak in most countries. Only a minority have explicit strategies linking climate and industrial policies to TVET reform and green-skills development; where initiatives exist, they are often sector-specific, time-bound and donor-dependent (ILO, 2019). Education sector plans, in particular, rarely articulate clear green-skills strategies (UNESCO, 2019).

This narrative aligns closely with UNESCO's ESD for 2030 Roadmap, which calls for "learning to live for our planet" and for national policies that green curricula, schools, teacher education and communities (UNESCO, 2020). It also foregrounds the role of higher education and TVET in research, innovation and professional formation for climate-relevant sectors.

The risk, however, is that a narrow focus on "green jobs" may instrumentalize education as a labor-market tool and sideline broader aims, citizenship, justice, critical thinking, unless these are intentionally integrated. As Kwauk and Casey (2021), Anderson (2012) and critical green-skills analyses argue, education for the green economy must remain anchored in a wider vision of sustainable development, human flourishing and transformation, not only in employability.

2.2.4 Education for Sustainable Development: Integrative and Holistic Approaches

A fourth narrative embeds climate change education within the broader project of Education for Sustainable Development (ESD), anchored in SDG 4.7. From this perspective, climate change is one dimension of interconnected crises involving poverty, inequality, biodiversity loss, health, energy and peace. Education's task is to help learners understand and navigate these interdependencies, and to foster the values, dispositions and competencies needed for sustainable, just futures. For example, the World Course, is an interdisciplinary K-12 curriculum aligned to learning standards drawn from the UN Sustainable Development Goals and the Universal Declaration of Human Rights, in which the study of climate and sustainability is integrated with the study of other development goals (Reimers et al., 2016).

UNESCO's ESD frameworks and global monitoring reports, together with Mochizuki and Bryan's conceptualization of Climate Change Education for Sustainable Development, converge on several pedagogical principles (UNESCO, 2017, 2020; Mochizuki & Bryan, 2015; Anderson, 2012):

- Cross-cutting integration. ESD should be woven through curricula rather than treated as a standalone subject; climate and sustainability themes should be integrated into science, social studies, citizenship, language, the arts and vocational subjects.
- Active, participatory learning. Learners should engage with real-world problems, local contexts and project-based activities rather than passively receiving information.
- Systems thinking. Students should be encouraged to see links between ecological, social and economic processes, and to understand feedback, trade-offs and synergies.
- Ethical reasoning and futures thinking. ESD should cultivate empathy, solidarity, respect for future generations and the ability to imagine alternative, more sustainable futures.
- Teacher development and whole-school approaches. Teachers and school leaders need sustained professional learning, not one-off workshops, to internalize ESD principles, adapt them to their subjects and lead whole-institution change (UNESCO, 2014, 2018, 2019).

UNESCO's analyses of curriculum and textbook content suggest that ESD themes have indeed become more visible over recent decades: references to environment and sustainability in textbooks and policy documents have increased, and many systems now claim to mainstream ESD and Global Citizenship Education (GCED) (UNESCO, 2016, 2019). At the same time, the Learn for our planet review concludes that the depth and balance of ESD-related learning remain limited. Across more than 260 official documents from 10 countries, references to ESD are heavily skewed towards cognitive learning; socio-emotional and behavioral dimensions receive far less emphasis (UNESCO, 2021). Similarly, in the 46-country review, environmental themes are present in most policies and curriculum frameworks but rarely elaborated in ways that specify action competences or socio-emotional outcomes (UNESCO, 2019, 2021).

Mochizuki and Bryan (2015) argue that CCESD should deliberately combine mitigation and adaptation dimensions, and that it should use climate change as an entry point for rethinking development trajectories, challenging inequality and building both resilience and responsibility. This requires explicit attention to climate justice, human rights and intergenerational equity, not only to technical solutions.

Implementation challenges are significant. ESD's breadth can easily become a liability: if "everything" is included, climate change can lose salience; if multiple goals compete, implementation can become diffuse and symbolic. UNESCO and ILO reviews highlight that many education sector plans and curriculum frameworks mention ESD and sustainability rhetorically but devote little instructional time, assessment weight or teacher-education attention to concrete climate and sustainability competencies (ILO, 2019; UNESCO, 2019, 2021).

Yet, there are promising practices. Whole-school approaches, where curriculum, pedagogy, school management, infrastructure and community links are aligned around sustainability, are associated with deeper student learning and more sustained behavior change (UNESCO, 2011, 2013; Hargis & McKenzie, 2020). Eco-school programs, green school certification schemes and climate-ready school guides illustrate how ESD can move beyond the classroom to transform institutional cultures and practices.

In sum, the ESD narrative offers a powerful integrative frame that situates climate education within broader struggles for sustainable, peaceful and equitable societies. Its effectiveness, however, depends on moving from declarative policy language to concrete, resourced strategies for teacher education, curriculum, assessment and whole-school change, and on ensuring that climate-specific knowledge and skills are not diluted in an overly broad agenda.

2.2.5 Critical and Decolonial Climate Education: Justice, Power and Plurality

The fifth narrative interrogates the epistemic and political underpinnings of mainstream CCE and ESD. It argues that many climate education initiatives, even when well-intentioned, reproduce "climate coloniality": they privilege Western scientific and technocratic perspectives, marginalize Indigenous and local knowledges, and downplay historical and ongoing processes of extraction, dispossession and structural violence that have produced climate vulnerability (Scherrer & Jules, 2025; Mbah & Ezegwu, 2024; Stein, 2019; Stein & Hare, 2023).

From this vantage point, education is not neutral: it is a terrain of struggle over whose knowledge counts, whose futures are imagined, and who bears responsibility for change. Decolonial CCE calls for:

- Recognizing colonialism and capitalism as root causes of the climate crisis.
- Centering the experiences, knowledge and priorities of frontline communities, particularly in the Global South.
- Critically examining power relations within education systems themselves (curricula, language policies, assessment regimes, teacher–student relations).
- Integrating Indigenous knowledge systems and place-based practices in ways that avoid tokenism.
- Experimenting with speculative and creative pedagogies ("engineering fiction," future-casting, arts-based methods) that allow learners to imagine alternative worlds.
- Linking CCE with peace education and post-conflict reconstruction, recognizing the entanglement of environmental degradation, violence and trauma.

UNESCO's Learn for our planet report and related scholarly reviews highlight that, despite rhetorical commitments, Indigenous knowledge remains weakly integrated in most curricula and teacher-education programs; where it appears, it is often inaccurate, de-contextualized or disconnected from sustainability priorities (Kim & Dionne, 2014; Whitehouse et al., 2014; UNESCO, 2021). Survey respondents report limited involvement of Indigenous stakeholders in school-based environmental activities, raising questions about whose perspectives and interests shape environmental learning.

In South Asia and elsewhere, Mbah et al. (2022) and related work argue for CCE that is explicitly oriented toward adaptation, justice and local agency. They propose that effective climate-adaptation education should:

- Be contextual and place-based, building on learners' attachments to local landscapes and cultures.
- Treat students and communities as co-creators of knowledge, not passive recipients of Northern expertise.
- Encourage critical examination of power structures and of whose voices are included or excluded in climate decisions.
- Connect reflection and analysis ("critical consciousness") with collective action ("praxis") to transform unjust conditions.

Kwauk and Casey (2021) extend this by calling for a "feminist planetary consciousness," an awareness of how the same logics that justify exploitation of the environment underpin exploitation of girls, women and marginalized groups. From this perspective, climate education must be gender-transformative: it should challenge patriarchal norms that restrict girls' access to climate-relevant learning and green jobs, surface how gendered divisions of labor shape climate vulnerability and adaptive capacity, and support girls' and women's leadership in climate decision-making.

This narrative also underscores that climate-literate, critically engaged citizens are better equipped to participate in democratic climate governance. Steves and Teytelboym (2013) find that public knowledge of climate change is a stronger predictor of effective climate-policy adoption than the formal level of democracy alone. Delina (2018) shows how climate mobilizations and democratic practices can be mutually reinforcing, especially in developing-country contexts.

The critical/decolonial narrative challenges educators and policy-makers to move beyond additive approaches ("including" Indigenous stories or a gender box in an otherwise unchanged curriculum) toward deeper transformations in how climate problems and solutions are framed. It overlaps with, but goes beyond, ESD and green-skills narratives by insisting that education attend to root causes, power relations and plural worldviews, and by foregrounding transformative capacities such as political agency, coalition-building, solidarity and systems-change skills that Kwauk and Casey (2021) identify as central to a "green learning agenda."

2.2.6 An Emerging Narrative, Climate Education as Part of AI Literacy

Education systems are rapidly incorporating artificial intelligence (AI) into curricula, particularly in response to the rise of generative AI tools. In parallel, definitions of

AI literacy have expanded beyond technical proficiency to include ethical reasoning, social responsibility, and applications for societal impact (Tadimalla & Maher, 2025). This evolution is worth noting for climate change education because it will influence the narrative of "skills for a sustainable future," which was explored earlier in this chapter.

While AI literacy has not yet emerged as a dominant narrative within climate change education, there are important convergences. The emerging AI literacy discourse increasingly embeds environmental sustainability and social accountability as core components, rather than peripheral concerns. For example, the AI Literacy (AILit) Framework for primary and secondary education, a joint initiative of the European Commission and the OECD, supported by Code.org and international experts, explicitly situates AI within broader societal and environmental contexts. The framework emphasizes that "values, context, and accountability are inseparable from learning with and about AI," and calls on learners to consider the environmental costs associated with AI systems, including energy use, material extraction, water consumption, and carbon emissions (European Commission & OECD, 2025).

Importantly, the framework does not treat environmental impact as an ancillary issue. Learners are asked to "describe how AI systems consume energy and natural resources" and to weigh the benefits of AI applications against their environmental and societal costs. A key knowledge competency highlights that AI systems "rely on vast but finite natural resources" and require human judgment to determine "whether the benefits of an AI tool is worth the environmental or societal cost." Skills and attitudes embedded in the framework further emphasize self and social awareness, responsible use, and empathy toward individuals, communities, and the environment (European Commission & OECD, 2025).

This integration of environmental reasoning into AI literacy signals an important trend for climate education. AI is simultaneously a rapidly expanding technological domain with significant environmental footprints; a tool increasingly used in climate modeling, mitigation planning, energy optimization, and adaptation; and a subject of ethical and political contestation regarding equity, access, and power. These dynamics create both tensions and opportunities. On one hand, AI literacy may compete for curricular space in already crowded programs, particularly where these digital skills are framed primarily as economic necessities. On the other hand, AI literacy could reinforce and deepen climate literacy by foregrounding systems thinking, ethical trade-offs, and the material consequences of digital infrastructures.

There is also emerging work, particularly in higher education and professional context that explores AI's applications for climate mitigation and adaptation, such as Climate Change AI, which is a nonprofit training and sharing knowledge of machine learning and AI applications for climate solutions (Climate Change AI, n.d.). However, this application-oriented dimension remains less visible in primary and secondary climate education frameworks. Bridging these domains could strengthen both fields: climate education can provide ethical and ecological grounding for AI use, while AI literacy can offer tools and critical lenses for analyzing and addressing climate challenges.

For now, AI literacy should not be understood as a replacement for climate literacy, nor as a new dominant narrative within climate change education. Rather, it represents an evolving adjacent discourse that may reshape how future skills are defined. As education systems negotiate the balance between digital transformation and ecological sustainability, the intersection of AI and climate literacies may become increasingly relevant.

2.3 Conclusion

Taken together, these five narratives illuminate the richness and complexity of second-order effects in the education–climate relationship. Education is simultaneously a site for building climate literacy, catalyzing behavior change, preparing green skills, advancing holistic sustainability, and contesting injustice.

However, global evidence also reveals significant disconnects between these ambitions and current practice. Environmental and climate themes remain marginal in many curricula and education sector plans, and where they appear they are heavily weighted towards cognitive outcomes; teacher education and assessment rarely prioritize socio-emotional or action competences; coordination between climate, education and skills policies is weak; and Indigenous, gender-transformative and decolonial perspectives are under-represented (ILO, 2019; Sabarwal et al., 2024; UNESCO, 2018, 2019, 2021).

As later chapters will show, concrete policies and programs rarely embody a single narrative; rather, they braid together elements of several, with different emphases depending on context. Kwauk and Casey's (2021) "new green learning agenda" is one attempt to articulate such a braid: combining specific skills for green jobs, generic green life skills and transformative capacities for systems change, and arguing that all three are needed if education is to respond adequately to both the technical and adaptive dimensions of the climate crisis. Chang's (2014, 2023) "knowing, doing and being" framework, Xu and Iyengar's (2023) systems-thinking primer, Lê's (2025) NGSS-aligned guide for secondary teachers and Worth's (2021) account of U.S. miseducation each, in different ways, exemplify how scholarship and practice are beginning to grapple with this complexity.

The task, then, is not to choose one narrative, but to understand how particular configurations of aims, content and pedagogy, and particular combinations of specific, generic and transformative capacities, can best serve learners and communities facing very different climate realities. This requires not only curricular and pedagogical innovation, but also policy coherence, investment in teacher and green-skills systems, and sustained dialogue across sectors, disciplines and knowledge traditions. The following chapter examines how national governments are attempting to translate these competing narratives into policy instruments and implementation strategies for climate-ready education systems.

References

ADB. (2021). *Getting education back on track: Partnership report*. Asian Development Bank. https://www.adb.org/multimedia/partnership-report2021/stories/getting-education-back-on-track/

Akita, T., & Miyata, S. (2021). Urban and rural dimensions of the role of education in inequality: A comparative analysis between Indonesia, Myanmar, and the Philippines. In *New frontiers in regional science: Asian perspectives* (pp. 127–166). Springer. https://doi.org/10.1007/978-981-16-1232-9_6

Alabi, B., Bonaga, J. K., Lin, M., & Page, N. (2026). Learning in crisis: Strengthening the Philippine education system. In F. M. Reimers, Z. Azim, M. A. Begi, J. Chia, & M. N. Shifat (Eds.). *Leading education systems: Learning from policy challenges and practices around the world*. Independently published.

Anderson, A. (2012). Climate change education for mitigation and adaptation. *Journal of Education for Sustainable Development, 6*(2), 191–206. https://doi.org/10.1177/0973408212475199

Angwas, R. O. (2025). School heads' emergency management competencies: Its influence on continuity of teaching and learning during emergency situations in the division of Agusan del Sur, Philippines. *International Journal of Scientific and Research Publications, 15*(6), 203–207. https://doi.org/10.29322/ijsrp.15.06.2025.p16225

Ariyasirichot, W. (2021). Policy design for preparation and coping with flash floods in the Mekong River basin: A case study of Mueang Nong Khai District, Thailand. *Turkish Journal of Computer and Mathematics Education, 12*(3), 4106–4112. https://doi.org/10.17762/turcomat.v12i3.1701

Aryal, J. P., Sapkota, T. B., Rahut, D. B., Marenya, P., & Stirling, C. M. (2021). Climate risks and adaptation strategies of farmers in East Africa and South Asia. *Scientific Reports, 11*, 10489. https://doi.org/10.1038/s41598-021-89992-5

ASEAN. (2015). *ASEAN safe schools initiative: A compilation of case studies*. Plan International. https://plan-international.org/uploads/sites/57/2021/12/asean_safe_schools_initiative_compilation_of_case_studies.pdf

Burke, S. E. L., Sanson, A. V., & Van Hoorn, J. (2018). The psychological effects of climate change on children. *Current Psychiatry Reports, 20*(5), 35. https://doi.org/10.1007/s11920-018-0896-9

Cas, A. G. (2016). Typhoon aid and development: The effects of typhoon-resistant schools and instructional resources on educational attainment in the Philippines. *Asian Development Review, 33*(1), 183–201. https://doi.org/10.1162/adev_a_00065

Chang, C. H. (2014). *Climate change education: Knowing, doing and being*. Routledge.

Chang, C. H. (2023). *Climate change education: Knowing, doing and being* (2nd ed.). Routledge.

Chang, C. H., & Pascua, L. (2015). Singapore students' misconceptions of climate change. *International Research in Geographical and Environmental Education, 25*(1), 84–96. https://doi.org/10.1080/10382046.2015.1106206

Chua Reyes, V., Hamid, O., & Hardy, I. (2021). When reforms make things worse: School leadership responses to poverty, disasters, and cultures of crises in the Philippine education system. *International Journal of Leadership in Education, 25*(2), 331–344. https://doi.org/10.1080/13603124.2021.2009038

Climate Change AI. (n.d.). *Climate Change AI*. https://www.climatechange.ai/

Comfort, L. K. (2012). Designing disaster resilience and public policy: Comparative perspectives, part II. *Journal of Comparative Policy Analysis, 14*(3), 199–201. https://doi.org/10.1080/13876988.2012.696440

Cooper, H., Valentine, J. C., Charlton, K., & Melson, A. (2003). The effects of modified school calendars on student achievement and school and community attitudes. *Review of Educational Research, 73*(1), 1–52. https://doi.org/10.3102/00346543073001001

Dare, E. A., Keratithamkul, K., Hiwatig, B. M., & Li, F. (2021). Beyond content: The role of STEM disciplines, realworld problems, 21st century skills, and STEM careers within science teachers' conceptions of integrated STEM education. *Education Sciences, 11*(11), 737. https://doi.org/10.3390/educsci11110737

Delina, L. L. (2018). *Climate actions: Transformative mechanisms for social mobilisation.* Palgrave Macmillan. https://doi.org/10.1007/978-3-319-91884-6

DepEd. (2024, May 16). *DepEd to build resilient schools, ensure recovery of disaster-affected infrastructures.* Department of Education, Republic of the Philippines. https://www.deped.gov.ph/2024/05/16/deped-to-build-resilient-schools-ensure-recovery-of-disaster-affected-infrastructures/

Deuchert, E., & Felfe, C. (2015). The tempest: Short- and long-term consequences of a natural disaster for children's development. *European Economic Review, 80*, 280–294.

EDCOM II, Senate of the Philippines, & House of Representatives of the Philippines. (2024). *Fixing the foundations: A matter of national survival (Year Two Report).* Second Congressional Commission on Education. https://edcom2.gov.ph/

European Commission & OECD. (2025). *AI literacy framework: Review draft.* https://ailiteracyframework.org/wp-content/uploads/2025/05/AILitFramework_ReviewDraft.pdf

Fagan, M., & Huang, C. (2019, April 18). A look at how people around the world view climate change. *Pew Research Center*. https://www.pewresearch.org/short-reads/2019/04/18/a-look-at-how-people-around-the-worldview-climate-change/

GPE (Global Partnership for Education). (2024). *Education for system transformation brief: Education resilience.* https://www.globalpartnership.org

Hargis, K., & McKenzie, M. (2020). Responding to climate change education: A primer for K–12 education. *Sustainability and Education Policy Network.* https://sepn.ca/resources/report-responding-to-climate-changeeducation-a-primer-for-k-12-education/

Herrera-Almanza, C., & Cas, A. G. (2020). Mitigation of long-term human capital losses from natural disasters: Evidence from the Philippines. *The World Bank Economic Review, 35*(2), 436–460. https://doi.org/10.1596/40807

Holmlund, T. D., Lesseig, K., & Slavit, D. (2018). Making sense of STEM education in K-12 contexts. *International Journal of STEM Education, 5*, Article 32. https://doi.org/10.1186/s40594-018-0127-2

ILO. (2019). *Skills for a greener future: A global view based on 32 country studies.* International Labour Organization.

Iuchi, K., Jibiki, Y., Solidum, R., & Santiago, R. (2019). *Natural hazards governance in the Philippines.* Oxford University Press. https://doi.org/10.1093/acrefore/9780199389407.013.233

Jones, C. A., & Davison, A. (2021). Disempowering emotions: The role of educational experiences in social responses to climate change. *Geoforum, 118*, 190–200. https://doi.org/10.1016/j.geoforum.2020.11.006

Kabir, M. H., Ahmed, I., Von Meding, J., Forino, G., & Ghosh, A. K. (2021). Impacts of prolonged waterlogging on educational continuity at schools in peri-urban Dhaka. *The Dhaka University Journal of Earth and Environmental Sciences, 9*(1), 13–23. https://doi.org/10.3329/dujees.v9i1.54857

Kim, E. J. A., & Dionne, L. (2014). Traditional ecological knowledge in science education and its integration in grades 7 and 8 Canadian science curriculum documents. *Canadian Journal of Science, Mathematics and Technology Education, 14*(4), 311–329. https://doi.org/10.1080/14926156.2014.970906

Kwauk, C., & Casey, O. (2021). *A new green learning agenda: Approaches to quality education for climate action.* Brookings Institution.

Lagmay, E. A. D., & Rodrigo, M. M. T. (2022). The impact of extreme weather on student online learning participation. *Research and Practice in Technology Enhanced Learning, 17*(1), 1–21. https://doi.org/10.1186/s41039-022-00201-2

Lambert, J. L., & Bleicher, R. E. (2014). Improving climate change communication starting with environmental educators. *Journal of Geoscience Education, 62*(3), 388–401. https://doi.org/10.5408/13-062.1

Lawson, D. F., Stevenson, K. T., Peterson, M. N., Carrier, S. J., Strnad, R. L., & Seekamp, E. (2019). Children can foster climate change concern among their parents. *Nature Climate Change, 9*, 458–462.

Leal Filho, W., Balasubramanian, M., Abeldaño, R. A., & Sierra, J. (2023). The effects of climate change on children's education attainment. *Sustainability, 15*(7), 6320.

Leiserowitz, A., Maibach, E. W., Roser-Renouf, C., Feinberg, G., & Howe, P. (2013, April 1). Climate change in the American mind: Americans' global warming beliefs and attitudes in April 2013 (Research report). *Yale Project on Climate Change Communication & George Mason University Center for Climate Change Communication.*

Lê, K. T. (2025). *Teaching climate change for grades 6–12: Activating science teachers to take on the climate crisis through NGSS* (2nd ed.). Routledge.

Maspul, K. A. (2024). Exploring STEM education for real-world climate change concerns to empower students as change agents. *Journal of Physics Education and Science, 1*(2), 1–12. https://doi.org/10.47134/physics.v1i2.249

Mbah, M. F., & Ezegwu, C. (2024). The decolonisation of climate change and environmental education in Africa. *Sustainability, 16*(9), 3744.

Mbah, M. F., Shingruf, A., & Molthan-Hill, P. (2022). Policies and practices of climate change education in South Asia: Towards a support framework for an impactful climate change adaptation. *Climate Action, 1*, 28. https://doi.org/10.1038/s44168-022-00029-0

Mochizuki, Y., & Bryan, A. (2015). Climate change education in the context of education for sustainable development: Rationale and principles. *Journal of Education for Sustainable Development, 9*(1), 4–26.

Monroe, M. C., Plate, R. R., Oxarart, A., Bowers, A., & Chaves, W. A. (2019). Identifying effective climate change education strategies: A systematic review of the research. *Environmental Education Research, 25*(6), 791–812.

Nassirpour, A., Galasso, C., & D'Ayala, D. (2018). Multi-hazard physical vulnerability prioritization of school infrastructure in the Philippines. In *Proceedings of the 11th U.S. National Conference on Earthquake Engineering.*

OECD. (2018). *Building resilient cities: An assessment of disaster risk management policies in Southeast Asia.* OECD Publishing. https://www.oecd.org

Opabola, E. A., & Galasso, C. (2024). Informing disaster-risk management policies for education infrastructure using scenario-based recovery analyses. *Nature Communications, 15*(1). https://doi.org/10.1038/s41467-023-42407-y

Pacific Disaster Center. (2021). *Philippines disaster management reference handbook.* https://reliefweb.int/report/philippines/2021-philippines-disaster-management-reference-handbook

Pal, A., Tsusaka, T. W., Nguyen, T. P., & Ahmad, M. M. (2023). Assessment of vulnerability and resilience of school education to climate-induced hazards: A review. *Development Studies Research, 10*(1). https://doi.org/10.1080/21665095.2023.2202826

Park, R. J., Behrer, A. P., & Goodman, J. (2021). Learning is inhibited by heat exposure, both internationally and within the United States. *Nature Human Behaviour, 5*(1), 19–27.

Pascua, L., & Chang, C. H. (2015). Using intervention-oriented evaluation to diagnose and correct students' persistent climate change misconceptions: A Singapore case study. *Evaluation and Program Planning, 52*, 70–77. https://doi.org/10.1016/j.evalprogplan.2015.04.001

Pew Research Center. (2019, February 10). Climate change still seen as the top global threat, but cyberattacks a rising concern. https://www.pewresearch.org/global/wp-content/uploads/sites/2/2019/02/Pew-Research-Center_Global-Threats-2018-Report_2019-02-10.pdf

Preña, E. M., & Labayo, C. P. (2024). Policy responses to extreme heat and its impact on education: The Philippine experience. *Policy Futures in Education, 23*(3), 652–675. https://doi.org/10.1177/14782103241288276

Prentice, C. M., Vergunst, F., Minor, K., & Berry, H. L. (2024). Education outcomes in the era of global climate change. *Nature Climate Change, 14*, 214–224.

Ramos, R. A. (2024). Schools as evacuation centers: Impacts on learning continuity in the Philippines. *Disaster Prevention and Management, 33*(2), 185–199.
Reimers, F., Chopra, V., Chung, C., Higdon, J., & O'Donnell, E. B. (2016). *Empowering global citizens: A world course.* CreateSpace Independent Publishing Platform.
Ring, E. A., Dare, E., Crotty, E. A., & Roehrig, G. (2017). The evolution of teacher conceptions of STEM education throughout an intensive professional development experience. *Journal of Science Teacher Education, 28*(5), 444–467. https://doi.org/10.1080/1046560X.2017.1356671
Roach, T., & Whitney, J. (2022). Heat and learning in elementary and middle school. *Education Economics, 30*(1), 29–46.
Rooney-Varga, J. N., Brisk, A. A., Adams, E., Shuldman, M., & Rath, K. (2014). Student media production to meet challenges in climate change science education. *Journal of Geoscience Education, 62*(4), 598–608. https://doi.org/10.5408/13-050.1
Sabarwal, S., Venegas Marin, S., Spivack, M., & Ambasz, D. (2024). *Choosing our future: Education for climate action.* World Bank.
Scherrer, B. D., & Jules, T. D. (2025). Rethinking climate change in education: From climate coloniality to decolonial educational ecologies in comparative and international education. *Current Issues in Comparative Education, 27*(1).
Stein, S. (2019). The ethical and ecological limits of sustainability: A decolonial approach to climate change in higher education. *Australian Journal of Environmental Education, 35*(3), 198–212. https://doi.org/10.1017/aee.2019.17
Stein, S., & Hare, J. (2023). The challenges of interrupting climate colonialism in higher education: Reflections on a university climate emergency plan. *Harvard Educational Review, 93*(3), 289–312. https://doi.org/10.17763/1943-5045-93.3.289
Stevenson, K. T., Peterson, M. N., Bondell, H. D., Moore, S. E., & Carrier, S. J. (2014). Overcoming skepticism with education: Interacting influences of worldview and climate change knowledge on perceived climate change risk among adolescents. *Climatic Change, 126*(3–4), 293–304. https://doi.org/10.1007/s10584-014-1228-7
Steves, F., & Teytelboym, A. (2013). *Political economy of climate change policy (SSEE Working Paper No. 13-02).* Smith School of Enterprise and the Environment.
Sprague, N. L., Scott, S. N., Mehranbod, C. A., Sachs, A. L., Ekenga, C. C., Rundle, A. G., Branas, C. C., & Factor-Litvak, P. (2025). Changing degrees: A weight of evidence scoping review examining the impact of childhood exposures to climate change on educational outcomes. *Environmental Research, 277*, Article 121639.
Taber, F., & Taylor, N. (2009). Climate of concern: A search for effective strategies for teaching children about global warming. *International Journal of Environmental & Science Education, 4*(2), 97–116.
Tadimalla, S. Y., & Maher, M. L. (2025). AI literacy as a core component of AI education. *AI Magazine, 46*, Article e70007. https://doi.org/10.1002/aaai.70007
Tong, T. M., Shaw, R., & Takeuchi, Y. (2012). Climate disaster resilience of the education sector in Central Vietnam. *Natural Hazards, 63*(2), 685–709. https://doi.org/10.1007/s11069-012-0178-5
UNDRR. (2019). *Critical infrastructure including schools, health facilities and roads threatened by climate crisis.* United Nations Office for Disaster Risk Reduction.
UNESCO. (2011). *Education for sustainable development: An expert review of processes and learning (ED.2010/WS/46).* UNESCO.
UNESCO. (2013). *Climate change education (Education Sector technical notes, ED/ADG/2013/01).* UNESCO.
UNESCO. (2014). *Shaping the future we want: UN Decade of Education for Sustainable Development (2005–2014), final report (ISBN 978-92-3-100053-9).* UNESCO.
UNESCO. (2018). *Education for sustainable development and global citizenship: Thematic indicators.* UNESCO.
UNESCO. (2019). *Country progress on climate change education, training and public awareness.* UNESCO.
UNESCO. (2020). *Education for Sustainable Development: A roadmap (ESD for 2030).* UNESCO.

UNESCO. (2021). *Learn for our planet: A global review of how environmental issues are integrated in education.* UNESCO.
UNESCO Global Education Monitoring Report Team. (2016). *Creating sustainable futures for all: Global education monitoring report, 2016; gender review* (56 pp.). UNESCO. https://doi.org/10.54676/JPTM4052
UNESCO MGIEP (Mahatma Gandhi Institute of Education for Peace and Sustainable Development). (2019). *Rethinking schooling for the 21st century: The state of education for peace, sustainable development and global citizenship in Asia.* (2nd ed.). UNESCO MGIEP.
UNICEF. (2024). *Learning interrupted: Global snapshot of climate-related school disruptions in 2024.* UNICEF. https://www.unicef.org/reports/learning-interrupted-global-snapshot-2024
UNICEF East Asia and Pacific Regional Office. (2025). *Under pressure: The impacts of climate change on education in East Asia and the Pacific.* UNICEF. https://www.unicef.org/eap
Whitehouse, H., Watkin Lui, F., Sellwood, J., Barrett, M. J., & Chigeza, P. (2014). Sea Country: Navigating Indigenous and colonial ontologies in Australian environmental education. *Environmental Education Research, 20*(1), 56–69. https://doi.org/10.1080/13504622.2013.852655
World Bank Group, & Asian Development Bank. (2021). *Climate risk country profile: Philippines.* World Bank; Asian Development Bank. https://openknowledge.worldbank.org
World Bank, UNICEF, FCDO, USAID, Bill & Melinda Gates Foundation, & UNESCO. (2022). *The state of global learning poverty: 2022 update.* World Bank.
World Health Organization. (2021). *COP26 special report on climate change and health: The health argument for climate action.* World Health Organization.
Worth, K. (2021). *Miseducation: How climate change is taught in America.* Columbia Global Reports.
Xu, C., & Iyengar, R. (2023). *Climate change education: An Earth Institute sustainability primer.* Columbia University Press.
You, J. (2024). Educational resilience in the face of climate change. *Lecture Notes in Education Psychology and Public Media, 71*(1), 157–163. https://doi.org/10.54254/2753-7048/2025.lc18316

Open Access This chapter is licensed under the terms of the Creative Commons Attribution-NonCommercial-NoDerivatives 4.0 International License (http://creativecommons.org/licenses/by-nc-nd/4.0/), which permits any noncommercial use, sharing, distribution and reproduction in any medium or format, as long as you give appropriate credit to the original author(s) and the source, provide a link to the Creative Commons license and indicate if you modified the licensed material. You do not have permission under this license to share adapted material derived from this chapter or parts of it.

The images or other third party material in this chapter are included in the chapter's Creative Commons license, unless indicated otherwise in a credit line to the material. If material is not included in the chapter's Creative Commons license and your intended use is not permitted by statutory regulation or exceeds the permitted use, you will need to obtain permission directly from the copyright holder.

Chapter 3
How Do National Governments Support Climate-Ready Education Systems?

This chapter examines how national governments are attempting to build climate-ready education systems through policy and implementation efforts focused on climate change education (CCE). Drawing on detailed country reviews for Italy, Scotland, Australia, Mexico, Indonesia and South Africa, complemented by the global analyses introduced in Chapters 1 and 2, we analyse the instruments, governance arrangements, and programmes that governments use to promote climate-change education in schools, and the persistent gaps between ambition and reality. We chose these countries because they have demonstrated and documented policies focusing on climate change education, and because they are in various continents and have diverse education systems in terms of financial resources and in terms of the education challenges they face. The selection is arbitrary and does not purport to be representative of other countries or regions, just to illustrate how countries adopting climate education policies do so. To write the reviews, we relied on publicly available information on government efforts in those countries to support climate change education in schools, the UNESCO-MECCE country reports, and published research studies.

We focus on second-order effects, how education systems respond to and act upon climate risks, rather than on first-order disruptions such as school closures or infrastructure damage. Following the UNESCO–MECCE and IBE "green education" frameworks (UNESCO, 2020, 2024a; UNESCO Global Education Monitoring Report Team & MECCE Project, 2024), we organise the analysis around four dimensions:

- Governance and legal mandates for CCE.
- Curriculum, teaching and assessment.
- Teacher education and professional development.
- Whole-school initiatives, partnerships and monitoring.

© The Author(s) 2026

F. Reimers and M. Wang-Aghania, *Educating for a Climate Changed Future*,
https://doi.org/10.1007/978-3-032-23256-4_3

Across the six countries, we see converging policy logics, anchored in environmental rights, education for sustainable development (ESD) and Action for Climate Empowerment (ACE), but also divergent trajectories in how these ideas are translated into curriculum, teacher education and school-level practice.

3.1 Governance and Legal Mandates

3.1.1 Climate Legislation that Names Education

In all six countries, climate change education is first signalled in climate-policy instruments rather than in education law per se, reflecting the ACE architecture.

Italy embeds education within a dense climate-policy framework: the National Strategy for Adaptation to Climate Change (2015) and the National Adaptation Plan (PNACC, 2023) emphasize awareness-raising, education and training as cross-cutting measures for adaptation (Ministry of the Environment, 2015; Ministry of Environment and Energy Security, 2023). The Decree-Law 111/2019 on climate emergency links climate action in schools to the EU Green Deal and SDGs, and supports subsequent education reforms (Italy, 2019b).

Mexico's General Climate Change Law (Federal Legislature, 2012, last amended 2024) assigns shared responsibilities for climate-change education and "climate culture" to the federation, states and municipalities, requiring them to promote education at all levels and public campaigns on climate causes, impacts and responses (SEMARNAT & INECC, 2018). The National Climate Change Strategy and the Special Climate Change Programme 2021–2024 further foreground environmental education as a tool for reducing vulnerability and building resilience (SEMARNAT, 2013, 2020).

Indonesia's climate strategy likewise calls for education-centered capacity building. The National Action Plan for Climate Change Adaptation (RAN-API) stresses the need to "socialize public awareness" and to provide "education, counselling and training" on climate adaptation, with the number of secondary-school learners educated about climate change an explicit indicator (Bappenas, 2014). The National Climate Change Learning Strategy (2013), prepared with UN CC: Learn, is explicitly subtitled "capacity building of human and institutional resources to address climate change," and has a dedicated section on formal education and teacher training (National Council on Climate Change, 2013).

South Africa's National Climate Change Response White Paper (Government of the Republic of South Africa, Department of Environmental Affairs, 2011) is among the most explicit in positioning education and training as strategic objectives, calling for mainstreaming climate knowledge into formal and non-formal education and linking climate-change education to green skills, public awareness and social justice (Government of the Republic of South Africa, Department of Environmental Affairs, 2011). The National Climate Change Adaptation Strategy (2019/2020) and the country's NDC reiterate education as an adaptation investment and an ACE obligation (Department of Environment, Forestry and Fisheries, 2019; Republic of South Africa, 2016).

Australia references climate education and communication across several climate and sustainability instruments. The National Climate Resilience and Adaptation Strategy 2021–2025 highlights the importance of climate information, awareness and "adaptation action and resilience building" in sectors including education (Australian Government, 2021a). The Long-Term Emissions Reduction Plan recognizes the role of higher education and skills training in supporting low-emissions technologies but offers limited detail on school-level climate-change education (Australian Government, 2021b).

Scotland, as a devolved administration within the UK, operates under UK-wide climate commitments but has differentiated goals, including a net-zero target for 2045 rather than 2050 (Scottish Government, 2019a). The Climate Change (Scotland) Act 2009 and its 2019 amendment, Scotland's contribution to the Paris Agreement and the Climate Change Plan Update 2018–2032 all frame public awareness, engagement and education as critical for emissions reduction and just transition (Scottish Government, 2019, 2020). The Net Zero Nation Public Engagement Strategy for Climate Change (2021) operationalises ACE pillars, "Understand, Participate, Act", and directly references learning for sustainability as part of public engagement (Scottish Government, 2021).

Across cases, climate-policy documents tend to treat education at a high level, in ACE-aligned terms of awareness, capacity and participation, without specifying content, time allocations or pedagogies for schools. These roles are more concretely defined in education-sector policies and curricula. By framing educational institutions as sites where communities access credible information and build adaptive capacity, these policies position schools as infrastructure for community-wide climate resilience. This extends their role beyond students to the wider public, a dynamic explored further in Chapter 8.

3.1.2 Education Laws and Sector Plans that Integrate Climate

A second layer of governance comprises education-sector statutes, curricula and sector plans that embed climate change within broader environmental or ESD mandates.

In Italy, Law 107/2015 ("La Buona Scuola") establishes sustainability and environmental education as objectives of school reform (Italy, 2015). A watershed development is Law 92/2019, which re-introduces civic education as a transversal, compulsory subject from primary to upper-secondary, mandating at least 33 hours per year and explicitly including sustainable development, environmental protection and climate change among its aims (Italy, 2019a). The Guidelines for Civic

Education (Ministerial Decree 35/2020) structure civics around three pillars, Constitution, sustainable development, digital citizenship, making education for sustainable development (ESD) and climate change a core curricular entitlement (Ministry of Education, 2020).

Scotland's Curriculum for Excellence (CfE), introduced from 2004, provides a 3–18 framework where Learning for Sustainability (LfS) is designated as a cross-curricular theme and entitlement. The Vision 2030 + report and the Learning for Sustainability Action Plan 2023–2030 ("Target 2030") commit to ensuring that by 2030 every learner experiences LfS and every 3–18 setting becomes a "Sustainable Learning Setting" (Scottish Government, 2023). LfS explicitly integrates sustainable development, global citizenship and outdoor learning, with climate change prominent in the sustainable development strand.

In Australia, environmental and sustainability education has been advanced through non-statutory but influential frameworks: Educating for a Sustainable Future (2005), Living Sustainably: The Australian Government's National Action Plan for Education for Sustainability (2009), and the Sustainability Curriculum Framework (2010) (Department of the Environment and Heritage, 2005; Department of the Environment, Water, Heritage and the Arts, 2009, 2010). These set out EfS principles and learning outcomes but lack legal force. The Australian Curriculum introduces Sustainability as a cross-curriculum priority (CCP) in Foundation–Year 10, describing climate change and biodiversity loss as contemporary issues students must be able to engage with (ACARA, n.d.-a). However, as discussed later, the CCP is advisory rather than mandatory, and climate change seldom appears as explicit content outside science and geography.

In Mexico, the 2019 reform of Article 3 of the Constitution and the new General Education Law (Gobierno de México, 2019) assign environmental education and climate resilience a central role. Articles 15, 16, 30 and 113 of the law require that study plans and programs integrate environmental education for sustainability, including prevention and reduction of climate change, and that SEP revise all curricula and materials to develop capacities in resource management, sustainable development and climate resilience (Gobierno de México, 2019; SEP, 2019). The Nueva Escuela Mexicana (NEM) pedagogical model further articulates respect for nature and climate-change prevention as one of eight guiding principles for basic education (SEP, 2019).

In Indonesia, Law 20/2003 on the National Education System requires curricula to consider regional environmental potential and links to global developments such as technology and climate change (Government of Indonesia, 2003). The 2013 National Curriculum Framework (MoECRT, 2013) includes climate change as a basic competence in Bahasa Indonesia at junior-secondary level, and the Standard Contents of Basic and Medium Education (MoECRT, 2016) specify that students in Years 7–12 should learn about global warming, anthropogenic causes and climate-related problem solving (MoECRT, 2013, 2016). The newer Merdeka (Independent) Curriculum reinforces climate integration through project-based learning under the Pancasila Student Profile (P5) theme of "sustainable lifestyle" (Tang, 2024; Teixeira & Crawford, 2022).

In South Africa, the White Paper on Education and Training (1995) established environmental education as a principle at all levels, later embedded in CAPS (Department of Education, 1995; DBE, 2011). CAPS situates environmental and climate concerns within principles of human rights, social justice and a healthy environment, with climate change appearing as a major content area in FET Geography and as a cross-cutting theme in Natural Sciences and Life Sciences (DBE, 2011; Thenga, 2025). The Action Plan to 2024 explicitly sets a strategic objective to improve education, awareness-raising and capacity on climate change mitigation and adaptation (DBE, 2020).

These education instruments differ in legal weight and specificity. Italy and Mexico introduce statutory requirements for climate- and sustainability-related content and hours; Scotland embeds LfS as an entitlement in curriculum and professional standards; South Africa incorporates environmental and climate themes in CAPS subject statements; Australia and Indonesia adopt sustainability/climate as cross-curricular priorities or basic competencies without enforceable time quotas. These differences reflect contrasting theories of change regarding whether climate education is best advanced through mandates, professional norms, or pedagogical flexibility.

3.2 Curriculum, Teaching and Assessment

Curriculum is the primary mechanism through which governments define what students should know, do and become in relation to climate change (Chang, 2014, 2022). Across the six cases, three broad curriculum integration models emerge:

- Dedicated civic/ESD subject with strong climate content (Italy).
- Cross-curricular sustainability frameworks embedded in a national curriculum (Scotland, Australia, Indonesia, Mexico, South Africa).
- Whole-subject anchoring with explicit climate content in specific disciplines (especially Geography and Sciences).

3.2.1 Dedicated Civic/ESD Subjects

Italy stands out for making climate-related ESD an explicit component of a compulsory, assessed subject. Law 92/2019 reinstates civic education with a minimum of 33 hours per year across all grades, and Ministerial Decree 35/2020's guidelines explicitly identify Education for Sustainable Development (ESD), including climate change, as one of three pillars (Ministry of Education, 2020). Attachments B and C specify climate-related learning outcomes by stage (e.g., understanding sustainable development and ecosystem respect by the end of lower-secondary; acting consistently with SDGs and eco-sustainable development by the end of upper-secondary).

This structures climate and sustainability learning as an entitlement for all students and creates a curricular "home" for climate-change education, while mandating cross-disciplinary links. However, this civic-education architecture evokes international praise but lacks systematic implementation monitoring. Teacher surveys and curriculum analyses suggest that high school autonomy leads to heterogeneous interpretation of civics and climate content, with some schools offering substantive climate projects and others limiting ESD to occasional activities (Antronico et al., 2023; Realdon et al., 2016).

3.2.2 *Cross-curricular Sustainability Frameworks*

Most systems rely on cross-curricular ESD or sustainability frameworks.

Scotland uses Learning for Sustainability as the principal carrier. LfS is not a standalone subject but a cross-cutting entitlement across CfE, connecting sustainable development, global citizenship and outdoor learning (Christie & Higgins, 2020). Climate change is one of the "defining issues of the twenty-first century" in Education Scotland materials and is woven into Sciences (Planet Earth), Social Studies, Technologies and Health and Wellbeing (Education Scotland, 2017a, 2017b, 2017c, 2019). The LfS Action Plan ("Target 2030") aims to ensure that every learner engages with LfS and that every setting becomes a sustainable learning environment by 2030 (Scottish Government, 2023).

Australia's Sustainability CCP is intended to "connect and relate relevant aspects of content across learning areas" (ACARA, n.d.-a). It explicitly lists climate change as a contemporary issue, but empirical analyses reveal that in practice climate references are sparse, often confined to later years and specific subjects. Gough (2021) found only ten sustainability-flagged statements across the F–10 Science curriculum, with little explicit mention of climate crisis. Eilam et al. (2020) showed that only 10 of 94 Victorian senior-secondary study designs contained climate-change references and none offered a complete conceptualization; some contained misconceptions (e.g., treating climate change as purely natural variability). Furthermore, Kennelly et al. (2011)and Nicholls and Thorne (2017/2018) argue that sustainability integration in the Australian Curriculum and teacher standards is rhetorically strong but lacks detailed guidance; teachers in Queensland showed strong support in principle but often low awareness of the Sustainability CCP and struggled to find time and resources to implement it.

Indonesia's Merdeka Curriculum retains the 2013 cross-curricular integration but enhances it through P5 project themes, including "sustainable lifestyle," giving schools flexibility to implement climate-related projects in science, social studies, language and extracurriculars (Nurhaliza et al., 2024; Teixeira & Crawford, 2022). However, Sofiyan et al. (2019) find a substantial gap between the proportion of geography basic competencies that are climate-related (50%) and the share of syllabus content that actually addresses them (22%), suggesting under-implementation. The basic-competency vs syllabus mismatch documented by Sofiyan et al. (2019) indicates that curricular frameworks offer opportunities for climate-integration, but official syllabi under-specify climate topics; teachers thus rely on improvisation. Studies of P5 implementation show that where project-based climate and environmental projects are well-resourced, they can significantly elevate climate literacy (e.g., Sagena et al., 2023; Salsabila, 2025).

Mexico's NEM promotes a transversal approach where climate- and sustainability themes are integrated across Spanish, Natural Sciences, Social Sciences and civic education, supported by free textbooks with climate content, especially in Grades 5–6 and lower-secondary (SEMARNAT & INECC, 2018). Curriculum attachments to the 2019 General Education Law require environmental education for sustainability and climate resilience but, as Paredes-Chi and Viga-de Alva (2018) note, environmental-education content in the 2009–2017 curriculum was conceptually diffuse and focused mainly on biophysical aspects, with limited attention to justice or action. Moreover, Eilam et al. (2020) and Paredes-Chi and Viga-de Alva (2018) highlight inconsistencies and omissions in how climate change is treated across subjects and levels. However, textbook analyses show both promise and limitations. Castañeda-Garza and Valerio-Ureña (2022) found that elementary textbooks contain 33 energy-literacy topics, often introduced earlier than expected by guidelines, suggesting a richer potential base for climate education than policy documents imply.

The cross-curricular model is consistent with ESD principles but, across cases, suffers from "everywhere and nowhere" risks: if climate and sustainability are everyone's responsibility but no one's core mandate, coverage can be inconsistent and shallow.

3.2.3 Whole-subject Anchoring with Explicit Climate Content

In contrast to cross-curricular sustainability frameworks, some systems anchor climate change more firmly within specific disciplines as a substantive unit of study. Across our cases, this approach is most clearly developed in South Africa, with partial exemplifications elsewhere (Mexico, Indonesia, Scotland).

South Africa's CAPS uses a topic-based infusion approach. In Senior Phase Social Sciences (Grades 7–9), topics on climate and natural environment provide building blocks; in FET Geography (Grades 10–12), climate change appears as a major unit encompassing causes, impacts, mitigation and adaptation (DBE, 2011; Thenga, 2025). However, implementation is left to teachers within prescribed time allocations. For example, Msezane (2017) documented variable alignment between CAPS environmental-impact content and exam coverage; some subjects, notably Life Sciences Agricultural Sciences, meet minimal allocations (3%) but offer little scope for deeper climate learning. Teachers and principals often lack a clear understanding of CAPS' environmental-learning aims (Damoah & Adu, 2020; Mokhele, 2011).

3.2.4 *Patterns Across Curriculum, Teaching, and Assessment*

Across all six case studies, curricular intent does not always translate into classroom reality, particularly when teachers face constraints in time, guidance, and resources. Mexico provides an interesting example: while climate content is integrated flexibly across subjects, it has influenced the development of textbooks and other learning materials that teachers are actively leveraging, demonstrating that flexibility, when paired with clear guidance and accessible resources, can support meaningful climate learning. This does not mean that cross-curricular or anchored approaches are ineffective; rather, their impact depends on systemic support, both formal, through aligned assessments and professional development (Chapter 6), and informal, through teacher communities and networks (Chapter 4).

Taken together, these studies echo UNESCO's global finding that climate-change content is increasingly present in curriculum frameworks but remains uneven, often superficial, and heavily cognitive (UNESCO, 2019, 2021, 2024a).

3.3 Teacher Education and Professional Development

As Chapters 1 and 2 emphasized, curriculum is not self-executing; climate-change education depends heavily on what teachers know, believe and are supported to do. Across our cases, governments use a mix of teacher-standards reforms, national networks and project-based PD to build teacher capacity.

3.3.1 *Professional Standards and Initial Teacher Education*

Some systems have begun to embed ESD/CCE in teacher-education standards:

In Scotland, the General Teaching Council for Scotland (GTCS) explicitly incorporates Learning for Sustainability into Professional Standards for Teachers. LfS is described as a cross-cutting theme underpinning teacher professionalism, and all teachers are expected to demonstrate commitment to sustainability and social justice, making LfS, and by extension, climate-change education, part of professional identity (GTCS, 2021).

In South Africa, the 2014 Minimum Requirements for Teacher Education Qualifications recognize sustainability as a foundational principle and situational competence for pre-service teachers (Council on Higher Education, 2014). Lotz-Sisitka (2015), however, notes that climate change is not named explicitly and that integration depends on individual teacher-educators.

In Italy, Law 92/2019 allocates €4 million annually from 2020 for teacher training on civic-education themes, including climate change, and requires updates to the National Teacher Training Plan; the Guidelines for Environmental Education and Sustainable Development (MIUR & MATTM, 2015) also specify teacher competences aligned with UNECE's ESD framework.

In Australia, Mexico and Indonesia, no national teacher-qualification standards explicitly require climate-change knowledge, though ESD or environmental education is sometimes referenced generically. In all cases, integration into ITE programs is patchy and heavily lecturer-dependent (e.g., Ferreira et al., 2007, 2009; Paredes-Chi & Viga-de Alva, 2018; Tang, 2025).

3.3.2 National and Regional PD Programs

Across countries, governments frequently partner with universities, NGOs and research institutes to deliver in-service climate-education PD.

Fundisa for Change (South Africa) is a national consortium (SANBI, DFFE, DBE, DHET, universities) offering SACE-endorsed PD modules on environment and sustainability aligned to CAPS, including specific units on climate change in Geography FET and Natural Sciences Senior Phase (Lotz-Sisitka, 2015). Studies show that Fundisa improves teachers' conceptual understanding and introduces more learner-centred pedagogies, but also that short PD interventions are insufficient to transform entrenched practices without ongoing support (Thenga et al., 2021).

Keep It Cool – Climate Change Education (South Africa) builds on Fundisa, training Geography and Natural Sciences teachers in three provinces and supporting school-based climate-action projects (VVOB, 2019). Mzilikazi et al. (2025) report that KIC-trained Natural Sciences teachers gained awareness and motivation but still lacked depth in climate science and practical strategies, emphasizing the need for longer-term, district-led PD.

Scotland's climate-education PD ecosystem includes Education Scotland's LfS guidance, GTCS resources, local authority programs and third-sector offerings such as Keep Scotland Beautiful's Climate Ready Classrooms (secondary and primary) and train-the-trainer workshops for teachers (Keep Scotland Beautiful, n.d.-a, 2025). Research indicates that while PD exists, many teachers still struggle to understand LfS and feel under-prepared for climate-justice and action-oriented teaching; Christie et al. (2019) and Ward et al. (2025) highlight professional-learning gaps and call for more systematic PD.

In Australia, teacher PD in climate and sustainability has largely been initiated by NGOs, research institutes and state programs: CSIRO's Sustainable Futures, AAEE's Sustainability in Schools portal, state Sustainable Schools initiatives, and EU-funded projects (e.g., I SEE, GIS4Schools) involving Italian and Australian partners (Baglivo & Conti, 2022; Tasquier et al., 2016). Evaluations consistently show positive impacts for participating teachers and students but note that participation is voluntary and small-scale, with limited national coordination (Grasso et al., 2017; Ugolini et al., 2012).

Indonesia's National Climate Change Learning Strategy explicitly identifies teachers as a priority group for capacity building and calls for strengthening teacher PD on climate change (National Council on Climate Change, 2013). The Climate Change Knowledge Center provides some resources, and BSKAP's Climate Change Education Guidebook (2024) targets teachers implementing the Merdeka Curriculum; yet studies find that many educators are unaware of these resources and rely on informal sources (Pitaloka et al., 2025; Tang, 2024). Localized PD initiatives, such as mentoring programs in Malang, Adiwiyata training, and projects like Triple C, can achieve high levels of teacher understanding and integration skills (Hermanto et al., 2025; Sagena et al., 2023), but remain the exception.

In Mexico, SEMARNAT's CECADESU and INECC run workshops, seminars and manual development (e.g., El cambio climático y mis derechos for lower-secondary) in partnership with SEP and UNICEF (INECC, 2018). The NEM framework emphasizes environmental education, and SEP provides digital training and materials, yet systematic national PD in climate-change education is limited, with teacher capacity heavily shaped by local projects and NGOs (e.g., Mario Molina Center, environmental-education networks) (Bello Benavides et al., 2022; Paredes-Chi & Viga-de Alva, 2020).

Italy's approach to teacher PD is similarly decentralized. Law 92/2019 earmarks funds, and MIUR & MATTM's, 2015 guidelines detail teacher competences; CNR, INGV and other research bodies provide didactic kits, lab visits and webinars (e.g., LaMMA meteorology/climate modules, ISPRA didactical kit) reaching thousands of teachers and students (Grasso et al., 2017; ISPRA, 2010). However, MECCE's profile and academic studies note that climate-specific PD remains relatively scarce and fragmented (UNESCO, 2021; Fioramonti et al., 2021).

Across all six countries, international and national reviews converge on a central point: teacher capacity is the critical bottleneck. Even where high-quality PD exists, it rarely reaches all teachers, is often short-term and project-based, and competes with other PD demands (Lotz-Sisitka, 2015; Christie et al., 2019; Tang, 2025; Pitaloka et al., 2025).

3.4 Whole-school Initiatives and Learning Environments

Beyond curriculum and teacher education, governments support CCE through whole-school programs and infrastructure initiatives that seek to green school operations and use them as learning environments.

Adiwiyata (Indonesia) is a national Green School certification program, led by KLHK and implemented with MoECRT and MoRA. It encourages schools to adopt environmentally friendly policies, integrate environmental education into curricula, involve communities and manage facilities sustainably. Adiwiyata is regulated by MoEF guidelines (e.g., Regulation No. 23/2022). Hermanto et al. (2025) and Wibowo et al. (2025) show that Adiwiyata schools often have better ecological literacy and environmental practices, but that students' climate knowledge and actions are still generally low-to-moderate, and that evaluation and teacher expertise remain weak points.

Eco-Schools (South Africa), run by WESSA with DBE's endorsement, mirrors the global Eco-Schools framework. It has reached thousands of schools and over a million learners, facilitating environmental projects and PD (Damoah & Adu, 2022; Rosenberg, 2008). Evaluations show improvements in school environments and learner behaviors but also highlight uneven quality and access; in some districts Eco-Schools is scarcely available or well-supported (Buthelezi, 2015).

Eco-Schools and Climate Action Schools (Scotland), coordinated by Keep Scotland Beautiful with government support, constitute a backbone of school-level climate- and sustainability work. KSB's Climate Action Schools framework clusters programs like Eco-Schools, Climate Ready Classrooms, live climate lessons and Climate Action Week, aligning them with national LfS and climate-policy aims (Keep Scotland Beautiful, n.d.-b, 2025). The Scottish Government provides funding for Eco-Schools expansion and for climate-engagement funds, recognizing these initiatives as vehicles for Target 2030 implementation.

AuSSI and state "Sustainable Schools" (Australia) began as a joint federal–state initiative and now operate primarily through state programs (e.g., ResourceSmart Schools in Victoria, Sustainable Schools NSW, Sustainable Schools WA, Green Adelaide). These adopt whole-school frameworks combining curriculum, operations, community partnerships and student action. Studies report substantial energy savings (up to 50% reductions in electricity in some Queensland schools), behavioral change and stronger student engagement where such programs are well-implemented (Purnell et al., 2004; Buchanan et al., 2016). However, as with other contexts, participation is selective, and there is no nationwide mandate or monitoring.

Escuela Verde (Mexico), coordinated by CECADESU, encourages school environmental management and education for sustainability; schools can be certified at national level. De León Rodríguez and Infante Bonfiglio (2014) found that certified schools often adopt waste, water and biodiversity actions, but that program depth and pedagogical integration vary, tending to prioritize operational changes over critical socio-environmental learning.

In Italy, the Piano RiGenerazione Scuola (2021–) represents a systemic attempt to support a "cultural and ecological transition" of schools via four pillars (knowledge, behaviors, infrastructure, opportunities). It funds school gardens, labs, green community networks and an annual "Week of Re-Generation," tied to COP narratives (AgenSIR, 2021; Bianchi, 2021). This plan moves beyond curriculum to treat schools as hubs of ecological practice and climate action, though implementation and evaluation remain emergent.

These programs align with UNESCO's whole-school and Green School standards (UNESCO, 2024b) and illustrate government efforts to move climate-change education from the classroom to the institutional and community level. Yet, across cases, they reach only a fraction of schools, rely heavily on external funding and champions, and lack robust, system-level evaluation.

3.5 Monitoring, Evaluation and Policy Learning

A striking cross-cutting feature is the relative absence of dedicated CCE monitoring indicators in national climate and education Monitoring and Evaluation systems.

Indonesia's RAN-API is one of the few instruments to specify an indicator on the number of secondary students educated about climate change, but public data on its implementation are limited (Bappenas, 2014). The National Climate Change Learning Strategy also outlines monitoring mechanisms (literature reviews, coordination meetings, reporting), but evidence of systematic use is scarce (National Council on Climate Change, 2013).

South Africa's National Climate Change Response Monitoring and Evaluation System Framework focuses on emissions, mitigation and finance, without education or communication indicators; however, DBE's Grade 12 diagnostic reports informally serve as a barometer of learners' understanding of climate-related concepts (Department of the Environment, Forestry and Fisheries, 2019).

In Mexico, Italy, Scotland and Australia, climate-education progress is usually inferred from policy and program reporting, participation rates (e.g., numbers of Eco-Schools, Adiwiyata or Green Schools), or ad hoc surveys and research studies, rather than from agreed national indicators for SDG 4.7 or 13.3 (UNESCO, 2019, 2021; MECCE Project, 2025).

Scotland's Target 2030 Action Plan explicitly commits to co-developing a core set of LfS indicators with learners and teachers by 2024 (Scottish Government, 2023), but at the time of the underlying MECCE profile such indicators were still under development.

This absence complicates policy learning and accountability: governments can report on high-level commitments but lack systematic evidence on the depth and equity of CCE implementation across schools.

3.6 Cross-country Patterns and Gaps

Synthesizing across the six cases, several patterns emerge in how governments support climate-ready education systems.

First, curriculum and policy intent do not automatically translate into practice. Even where climate- and sustainability-related content is compulsory and embedded across subjects (Italy, Scotland, and South Africa), implementation varies widely depending on teacher capacity, time, and resources. Cross-curricular integration offers promise, but carries pitfalls. Embedding climate education across subjects aligns with ESD and ACE principles, but without clear guidance, assessment alignment, or structured time allocations, coverage can be superficial, primarily cognitive, and inconsistently applied. Dedicated subjects, whole-subject anchoring, or explicit teacher standards can partially mitigate these risks by giving climate content a clearer "home," yet autonomy, exam pressures, and uneven PD often limit their impact.

Mexico provides a case study in which clearer specification of where climate content sits within subjects has catalyzed textbook and resource development. However, focusing primarily on curricular content, especially when implementation emphasizes cognitive outcomes alone, can sideline socio-emotional development and action competences, key elements of climate literacy (UNESCO, 2021; Mochizuki & Bryan, 2015).

Second, teacher capacity is a critical bottleneck. Across all countries, teachers often lack deep understanding of climate change, and many feel unprepared to translate new content into practice (Anyanwu et al., 2015; Tang, 2025; Pitaloka et al., 2025; Mzilikazi et al., 2025; Christie et al., 2019; Ward et al., 2025; Nurhaliza et al., 2024). Short, project-based PD can raise awareness and initial knowledge, but lasting change appears to require embedded professional-learning systems, ongoing mentoring, PLCs, district-level support, as Scotland's and South Africa's emerging blended models are beginning to explore (GTCS, 2021; VVOB, 2024). The next chapter examines how informal teacher networks and communities of practice can complement formal systems, providing scalable, sustained avenues for professional development.

Third, reach and equity remain persistent concerns. Whole-school programs initiatives, such as Adiwiyata, Eco-Schools, AuSSI, and Italy's RiGenerazione, are more prevalent in better-resourced schools and provinces (Damoah & Adu, 2022; Rosenberg, 2008). In Indonesia and Mexico, rural, outer-island and Indigenous communities often have less access to PD, digital resources and well-supported programs, despite being highly climate-vulnerable (Satyawan & Baskara, 2023; Sandoval-Rivera & Mendoza-Zuany, 2021). These disparities echo UNESCO's global findings that high-quality climate-change education is unevenly distributed, often privileging urban and high-resource schools (UNESCO, 2021, 2024a).

Finally, in all six countries, CCE-specific monitoring is weak. Without agreed indicators for SDG 4.7/13.3 at national level, governments have limited visibility into where and how climate-change education is actually happening, and for whom. Education sector plans and diagnostic reports provide clues (e.g., South Africa's NSC diagnostics; Scotland's future LfS indicators), but none of the six has yet developed a comprehensive CCE monitoring framework linked to climate and education M&E systems.

3.7 Implications for Climate-Ready Education Systems

The six cases illustrate how national governments are beginning to operationalize the global ACE and ESD agendas in distinct ways:

- Italy and Mexico exemplify a legalistic, civics-centred route, using constitutional/education-law reforms and dedicated civic or environmental-education components to anchor climate-related learning.

- Scotland and South Africa illustrate a curriculum-plus-teacher-development route, embedding climate and sustainability across curriculum principles and subject content, while mobilizing teacher-education networks (LfS, Fundisa, KIC-CCE) to build capacity.
- Australia and Indonesia show a cross-curriculum-priority and Green School route, where sustainability/climate priorities and programs like AuSSI and Adiwiyata sit alongside a national curriculum but rely heavily on voluntary uptake and external partnerships.

All three pathways are compatible with UNESCO's Greening Education Partnership and IBE's Green Education for Sustainable Societies dimensions (UNESCO, 2020, 2024a; UNESCO IBE, 2022, 2023). Yet, across all models, moving from isolated efforts to climate-ready systems will require:

- Stronger alignment of climate-policy goals with education strategies, curricula, teacher standards and assessment.
- Substantial investment in teacher-education and professional-learning systems that treat climate change as a central, not peripheral, domain of practice.
- Robust monitoring and evaluation of climate-change education across regions, subjects and learner groups.
- Attention to equity and justice, ensuring that learners in the most climate-vulnerable communities have access to the richest climate-education opportunities.
- Whole-school approaches that align infrastructure, governance and community engagement with instructional CCE, modelling the "knowing, doing and being" that climate-ready education seeks to cultivate (Chang, 2014, 2022).

The six national experiences reviewed here show that governments are not starting from scratch: they have built significant policy scaffolding and experimented with promising programs. The central challenge, echoed across MECCE country profiles and empirical studies, is to convert this scaffolding into coherent, scaled, and equitable practice that can help current and future generations navigate, and transform, a rapidly warming world.

References

ACARA. (n.d.-a). *Sustainability*. Australian Curriculum, Assessment and Reporting Authority.

AgenSIR. (2021, December 28). Scuola: min. Istruzione, stanziati 102 milioni di fondi Pon su transizione ecologica ed educazione ambientale.

Antronico, L., Coscarelli, R., Gariano, S. L., & Salvati, P. (2023). Perception of climate change and geo-hydrological risk among high-school students: A local-scale study in Italy. *International Journal of Disaster Risk Reduction, 90*, Article 103663.

Anyanwu, R., Le Grange, L., & Beets, P. (2015). Climate change science: The literacy of Geography teachers in the Western Cape Province, South Africa. *South African Journal of Education, 35*(3), 1–9.

Australian Government. (2021a). *National climate resilience and adaptation strategy 2021–2025*. Department of Agriculture, Water and the Environment

Australian Government. (2021b). *Australia's long term emissions reduction plan*. Department of Industry, Science, Energy and Resources.

Baglivo, L., & Conti, D. (2022). Change the Story, learning to make a difference in climate crisis. *Environmental Sciences Proceedings, 14*(1), 4. https://doi.org/10.3390/environsciproc2022014004

Bappenas. (2014). Rencana Aksi Nasional Adaptasi Perubahan Iklim (RAN-API).

Bello Benavides, L. O., Sánchez, G., & Meza Mesa Ortiz, S. L. (2022). Project-based learning. In *Advances in game-based learning*.

Bianchi, G. (2021). *Il Piano "RiGenerazione Scuola": La transizione ecologica e culturale parte dalla scuola*. Istruzione Veneto.

Buchanan, J., Schuck, S., & Aubusson, P. (2016). In school sustainability action: Climate Clever Energy Savers. *Australian Journal of Environmental Education, 32*(2), 154–173. https://doi.org/10.1017/aee.2015.55

Buthelezi, S. S. (2015). *Assessment of the availability and effectiveness of environmental education through the Eco Schools Programme at Umzinyathi District schools* [Master's thesis]. University of KwaZulu Natal.

Castañeda-Garza, G., & Valerio-Ureña, G. (2022). Energy literacy in elementary school textbooks in Mexico. *Environmental Education Research, 28*(4), 562–585.

Chang, C. H. (2014). *Climate change education: Knowing, doing and being*. Routledge.

Chang, C. H. (2022). *Climate change education: Knowing, doing and being*. (2nd ed.). Routledge.

Christie, B., & Higgins, P. (2020). *Educational outcomes of Learning for Sustainability: A brief review of literature*. Scottish Government.

Christie, B., Higgins, P., King, B., Collacott, M., Kirk, K., & Smith, H. (2019). From rhetoric to reality: Examining the policy vision and the professional process of enacting Learning for Sustainability in Scottish schools. *Scottish Educational Review, 51*(1), 44–56.

Climate Change Knowledge Center, Rumah Belajar, & BSKAP. (2024). *Climate change education guidebook*.

Council on Higher Education. (2014). *Policy on the minimum requirements for teacher education qualifications*.

Damoah, B., & Adu, E. (2022). Environmental education in South African schools: The role of civil society organizations. *Research in Social Sciences and Technology, 7*(3), 1–17. https://doi.org/10.46303/ressat.2022.14

Damoah, B., & Omodan, B. (2022). Determinants of effective environmental education policy in South African schools. *International Journal of Educational Research Open, 3*, Article 100213.

De León Rodríguez, A. P., & Infante Bonfiglio, J. M. (2015). Una evaluación crítica de una experiencia de educación ambiental para la sustentabilidad en el nivel educativo básico en Nuevo León, México. CPU-e, *Revista de Investigación Educativa, (19)*, 184–212. https://doi.org/10.25009/cpue.v0i19.970

Department of Basic Education. (2011). *Curriculum and Assessment Policy Statements (CAPS)*.

Department of Basic Education. (2020). *Action plan to 2024: Towards the realisation of Schooling 2030*.

Department of Education. (1995). *White paper on education and training*. Government Gazette, 357(16312).

Department of Environment, Forestry and Fisheries. (2019). *National climate change adaptation strategy: Republic of South Africa*. DFFE.

Department of Forestry, Fisheries and the Environment. (2018). *South Africa's third national communication under the United Nations Framework Convention on Climate Change*. DFFE.

Department of the Environment, Water, Heritage and the Arts. (2009). *Living sustainably: The Australian Government's national action plan for education for sustainability*. Commonwealth of Australia.

Department of the Environment, Water, Heritage and the Arts. (2010). *Sustainability curriculum framework: A guide for curriculum developers and policy makers.* Commonwealth of Australia
Department of Environmental Affairs. (2011). *National climate change response white paper.*
Department of the Environment and Heritage. (2005). *Educating for a sustainable future: A national environmental education statement for Australian schools.* Commonwealth of Australia
Education Scotland. (2017a). Curriculum for excellence: Experiences and outcomes.
Education Scotland. (2017b). Opening up great learning – Learning for sustainability.
Education Scotland. (2017c). Weather and climate change.
Education Scotland. (2019). What is curriculum for excellence?
Eilam, E., Prasad, V., & Widdop Quinton, H. (2020). Climate change education: Mapping the nature of climate change, the content knowledge and examination of enactment in upper secondary Victorian curriculum. *Sustainability, 12*(2), 591.
Federal Legislature. (2012). General law on climate change [Ley General de Cambio Climático]. Diario Oficial de la Federación, Mexico.
Ferreira, J.-A., Ryan, L., & Tilbury, D. (2007). Mainstreaming education for sustainable development in pre service teacher education in Australia: A review of existing professional development models. *Australian Research Institute in Education for Sustainability*, Macquarie University.
Ferreira, J.-A., Ryan, L., Davis, J., Cavanagh, M., & Thomas, J. (2009). Mainstreaming sustainability into pre service teacher education in Australia. *Environmental Education Research, 15*(5), 607–623.
Fioramonti, L., Giordano, C., & Basile, F. L. (2021). Fostering academic interdisciplinarity: Italy's pioneering experiment on sustainability education in schools and universities. *Frontiers in Sustainability, 2*, Article 631610.
Grasso, V., Bartolini, G., Benedetti, R., Betti, G., Capecchi, V., Gozzini, B., Magno, R., et al. (2017). Engaging students and teachers in meteorology and atmospheric sciences: The LaMMA activities. *Advances in Science and Research, 14*, 109–114.
GTCS. (2021). *Professional standards for teachers.* General Teaching Council for Scotland.
Gobierno de México. (2019). Decreto por el que se reforman y adicionan diversas disposiciones de los artículos 3o., 31 y 73 de la Constitución Política de los Estados Unidos Mexicanos, en materia educativa. *Diario Oficial de la Federación.*
Gough, A. (2021). All STEM-Ed up: Gaps and silences around ecological education in Australia. *Sustainability, 13*(7), 3801.
Government of Indonesia. Law No. 20 of 2003 on the National Education System (Indonesia).
Government of the Republic of South Africa, Department of Environmental Affairs. (2011, October). *National climate change response white paper.* Government of the Republic of South Africa. https://www.gov.za/sites/default/files/gcis_document/201409/nationalclimatechangeresponsewhitepaper0.pdf
Hermanto, F., Lukitawati, L., Rusdini, S. E., & Amrullah, M. F. (2025). Reinventing ecopedagogy and its implication on enhancing climate action in Indonesian Adiwiyata school. *Asian Journal of Education and Social Studies, 48*(2), 112–130.
INECC. (2018). *Sexta Comunicación Nacional de México ante la CMNUCC.*
ISPRA – Italian National Institute for Environmental Protection and Research. (2010). *Environmental data yearbook 2010.* ISPRA.
Italy. (2015). Legge 13 luglio 2015, n. 107.
Italy. (2019a). Legge 20 agosto 2019, n. 92.
Italy. (2019b). Decreto-legge 14 ottobre 2019, n. 111.
Keep Scotland Beautiful. (n.d.-a). *Climate ready classrooms.*
Keep Scotland Beautiful. (n.d.-b). *Climate action schools.*
Keep Scotland Beautiful. (2025). *Climate education at scale – Is your class climate ready?*
KLHK. (2020). *Rencana Strategis Direktorat Jenderal Pengendalian Perubahan Iklim 2020–2024.*
Kennelly, J., Taylor, N., & Serow, P. (2011). Education for sustainability and the Australian Curriculum. *Australian Journal of Environmental Education, 27*(2), 209–218.

Lotz-Sisitka, H. (2015). Teacher professional development with an education for sustainable development focus in South Africa: Development of a network, curriculum framework and resources for teacher education. In *Empowering educators for a sustainable future* (pp. 111–134). UNECE.

MECCE Project. (2025). *Country profiles: Italy, Scotland, Australia, Mexico, Indonesia, South Africa.*

MIUR & MATTM. (2015). *Linee guida per l'educazione ambientale e allo sviluppo sostenibile.*

Ministry of Education. (2020). Decreto ministeriale 22 giugno 2020, n. 35.

Ministry of Environment and Energy Security. (2023). *Piano nazionale di adattamento ai cambiamenti climatici.*

Ministry of the Environment. (2015). *Strategia nazionale di adattamento ai cambiamenti climatici.*

Mochizuki, Y., & Bryan, A. (2015). Climate change education in the context of education for sustainable development: Rationale and principles. *Journal of Education for Sustainable Development, 9*(1), 4–26.

MoECRT. (2013). Kurikulum 2013: Kompetensi dasar sekolah menengah pertama. *Ministry of Education and Culture.*

MoECRT. (2016). Standar isi pendidikan dasar dan menengah. *Ministry of Education and Culture.*

Mokhele, M. L. (2011). Integrated environmental teaching in South Africa: An impossible dream? *South African Journal of Education, 31*(1), 37–48.

Msezane, S. B. (2017). *An analysis of the policy coverage and examination of environmental impact topics* [Master's thesis]. University of the Witwatersrand.

Mzilikazi, B., Khalo, X., & Caga, N. (2025). Understanding of natural sciences teachers on addressing climate change concepts in the curriculum: Evidence from Fundisa for Change Keep It Cool Climate Change Project. *Interdisciplinary Journal of Education Research, 7*(1), Article a11.

National Council on Climate Change. (2013). *Indonesia: National Climate Change learning strategy.*

Nurhaliza, S., Silvhiany, S., & Inderawati, R. (2024). Integrating climate change education in English lessons and P5 projects in junior high schools. *Journal of Languages and Language Teaching, 12*(1), 10–24.

Paredes-Chi, A., & Viga-de Alva, M. D. (2018). Environmental education policy and content of the contemporary (2009–2017) Mexican national curriculum for primary schools. *Environmental Education Research, 24*(5), 608–624.

Paredes-Chi, A., & Viga-de Alva, M. D. (2020). Participatory action research and environmental education: A Mexican experience with teachers from a primary rural school. *Environmental Education Research, 26*(8), 1115–1134.

Pitaloka, T. D. A., et al. (2025). Climate change education challenges in Indonesia: Educators' perspective. *Environment Education and Conservation, 10*(1), 1–15.

PNACC Ministry of Environment and Energy Security. (2023). Piano nazionale di adattamento ai cambiamenti climatici (PNACC). *Ministero dell'Ambiente e della Sicurezza Energetica.*

Purnell, K., Sinclair, M., & Gralton, A. (2004). Sustainable schools: Making energy efficiency a lifestyle priority. *Australian Journal of Environmental Education, 20*(2), 81–91.

Realdon, G., Paris, E., & Invernizzi, M. C. (2016). Teaching Earth Sciences in Italian liceo high schools following the 2010 reform: A survey. *Rendiconti Online della Società Geologica Italiana, 40*, 71–79.

Republic of South Africa. (2016). *South Africa's first nationally determined contribution (NDC).*

Rosenberg, E. (2008). Eco-Schools and the quality of education in South Africa: Realising the potential. *Southern African Journal of Environmental Education, 25*, 25–39.

Sagena, U. W., Azizsyah, S. N., Panjaitan, O., Masjaya, M., & colleagues. (2023). Program Triple C (Climate Change Class) untuk peningkatan kesadaran lingkungan bagi anak. *Jurnal Masyarakat Mandiri, 7*(2), 115–130.

Salsabila, I. N. (2025). Green Education Movement: Integrating environmental education in the curriculum to address the global climate crisis. *International Journal of Social Research, 5*(1), 50–70.

Sandoval-Rivera, J. C. A., & Mendoza-Zuany, R. G. (2021). The milpa goes to school: Teacher training in climate change adaptation from a situated learning perspective. *Sustainability and Climate Change, 14*(2), 87–98.

Satyawan, I. A., & Baskara, S. H. (2023). Climate change learning at junior and senior high schools on the outer islands of Indonesia: The case on Sebatik Island. *IOP Conference Series: Earth and Environmental Science, 1201*, Article 012011.

Scottish Government. (2019). *Climate Change (Emissions Reduction Targets) (Scotland) Act 2019: Policy memorandum.*

Scottish Government. (2020). *Update to the Climate Change Plan 2018–2032: Securing a green recovery on a path to net zero.*

Scottish Government. (2021). *Net Zero Nation: Public engagement strategy for climate change.*

Scottish Government. (2023). *"Target 2030": A movement for people, planet and prosperity – Scotland's Learning for Sustainability Action Plan 2023–2030.*

SEMARNAT. (2013). Estrategia Nacional de Cambio Climático. Visiön 10–20–40.

SEMARNAT. (2020). *Programa Sectorial de Medio Ambiente y Recursos Naturales 2020–2024.*

SEMARNAT & INECC. (2018). *Sexta Comunicación Nacional.*

SEP. (2019). *La Nueva Escuela Mexicana: Principios y orientaciones pedagógicas.*

Sofiyan, S., Aksa, F. I., & Saiman, S. (2019). An analysis of climate change of the curriculum in Indonesia. *Journal of Physics: Conference Series, 1317*, Article 012191.

Tang, K. (2024). Climate change education in Indonesia's formal education: A policy analysis. *Npj Climate Action, 3*, 12.

Tang, K. (2025). Teacher conceptions of climate change and their role in climate change education: Insights from Indonesian upper-secondary teachers. *Discover Sustainability, 6*(1), 45.

Tasquier, G., Levrini, O., & Dillon, J. (2016). Exploring students' epistemological knowledge of models and modelling in science: Results from a teaching/learning experience on climate change. *International Journal of Science Education, 38*(4), 539–563.

Teixeira, J., & Crawford, E. O. (2022). Climate change education and curriculum revision. In E. O. Crawford & A. Kirby (Eds.), *Climate change education: International perspectives* (pp. 213–233). Routledge.

Thenga, M. (2025). Barriers and opportunities in implementing climate change education in the FET Phase Geography curriculum in South Africa. *Journal of Geography Education in Africa, 8*, 1–15.

Thenga, M., Goldschagg, P., Ferguson, R., & Mandikonza, C. (2021). Teacher professional development and Geography teachers' pedagogical practices for climate change education. *Southern African Journal of Environmental Education, 37*, 64–83.

Ugolini, F., Massetti, L., Tagliaferri, G., Lanini, G. M., Screti, C., & Raschi, A. (2012). Introducing climate change topics in science teaching through the 5E model of IBL: Teachers' perspectives. In *Proceedings of an international conference on inquiry based science education.*

UNESCO. (2019). *Country progress on climate change education, training and public awareness.*

UNESCO. (2020). Education for Sustainable Development: A roadmap (ESD for 2030).

UNESCO. (2021). *Learn for our planet: A global review of how environmental issues are integrated in education.*

UNESCO. (2024a). *Greening every curriculum: Guidance to teach and learn for climate action.*

UNESCO. (2024b). *Green school quality standard: Greening every learning environment.*

UNESCO Global Education Monitoring Report Team, & MECCE Project. (2024). *Education and climate change: Learning to act for people and planet*. UNESCO.

UNESCO International Bureau of Education. (2022). *Transforming curriculum and learning for climate action.*

UNESCO International Bureau of Education. (2023). *Greening education systems: A curriculum centred approach.*

VVOB. (2019). *Keep It Cool: Climate change education project.*

VVOB. (2024). *Keep It Cool 2.0: Climate change education.*

Ward, K. S., Purcell, M., Dey, M. B., et al. (2025). Learning for sustainability in Scotland: When best practice is not enough. *Environmental Education Research, 31*(6), 1099–1116.
Wibowo, N. A., Sumarmi, S., Putra, A., Silviariza, W. Y., Yembuu, B., et al. (2025). Green education in addressing climate change: An empirical study of ecological literacy through the Adiwiyata program. *E3S Web of Conferences, 400*, 09003.

Open Access This chapter is licensed under the terms of the Creative Commons Attribution-NonCommercial-NoDerivatives 4.0 International License (http://creativecommons.org/licenses/by-nc-nd/4.0/), which permits any noncommercial use, sharing, distribution and reproduction in any medium or format, as long as you give appropriate credit to the original author(s) and the source, provide a link to the Creative Commons license and indicate if you modified the licensed material. You do not have permission under this license to share adapted material derived from this chapter or parts of it.

The images or other third party material in this chapter are included in the chapter's Creative Commons license, unless indicated otherwise in a credit line to the material. If material is not included in the chapter's Creative Commons license and your intended use is not permitted by statutory regulation or exceeds the permitted use, you will need to obtain permission directly from the copyright holder.

Chapter 4
Beyond Government: The Role of Networks Supporting Climate Change

Beyond the role that government policy plays in advancing climate change education in public school systems, education networks play a critical role in supporting and scaling such efforts, particularly in an increasingly digitized and globally connected world. While government-supported, systems-wide reform depends on policy coherence, support, and capacity which will be discussed Chap. 7, much progress in climate education currently sits outside of formal government mandates. An example of this are the school-based cases that will be discussed in Chap. 5, of which many are private school initiatives, and many among the rest reflect grassroot initiatives, unrelated to government mandates. The 'World's Best Schools Prizes' that recognize such exemplary practices are a private initiative of the non governmental organization T4Education. In this chapter, we examine initiatives where teachers experiment, learn from one another, and develop the professional capacity to build climate literacy among students which result from non-governmental initiatives. We will examine efforts of civil society involving networks of teachers and schools, some of them working within nations, and others across nations.

In a study of historical efforts to transform education in the United States, Cohen and Mehta (2017) emphasize the importance of structural support and aligned incentives for system-wide reforms to succeed. Their analysis is in line with the policy coherence framework discussed in Chaps. 1 and 7. Yet, they also argue that in the absence of system-wide support, reforms can still succeed as "niche reforms," initiatives that "create a space in the education sector by defining a novel and specialized mission and mobilizing resources to support and sustain it" (Cohen & Mehta, 2017). These niches function as protected environments in which new practices can be developed and demonstrated over time through subsystem-level infrastructure and support. In the case of climate education, this subsystem-level infrastructure comprises motivated educators, students, and community members. Specifically, these innovations are emerging and spreading from networks of educators who share resources, co-develop practices, and collectively build expertise in contexts where formal governmental guidance and support are limited.

© The Author(s) 2026

F. Reimers and M. Wang-Aghania, *Educating for a Climate Changed Future*,
https://doi.org/10.1007/978-3-032-23256-4_4

This chapter, therefore, focuses on networks of educators as a critical mechanism and bottom-up approach for building climate education capacity and scaling climate education. We will examine four types of networks: school- or district-based professional learning communities, international and regional teacher networks, digital & online networks, and research-educator partnerships. This chapter reframes informal networks of teachers as essential infrastructure for eventual system-wide reform particularly in contexts where political or institutional constraints limit government leadership. Even in contexts where there is strong political will and institutional support, activating networks of educators remains critical to scaling climate education.

4.1 Why Informal Teacher Networks Matter

Social network theory and networked learning research suggests that professional learning can occur outside formal professional development structures (Coutinho & Lisboa, 2013; Vaessen et al., 2014). Informal learning, characterized as unstructured, self-directed, experimental, and embedded in day-to-day practice, relies heavily on social interaction rather than institutionalized training (Ringling et al., 2021). For example, in a case study of school leaders, informal learning opportunities emerged as just as important as formal training since learning responded to immediate needs and was therefore, timely, situation-specific, and immediately applicable (Ringling et al., 2021). These interactions often occur frequently and spontaneously, reflecting the demand-driven nature of informal learning. Learners move fluidly between roles as knowledge seekers and knowledge holders, reinforcing reciprocal learning and professional agency.

Social network theory helps explain why these informal spaces matter: knowledge is not simply transmitted through formal channels, but created and exchanged through relationships among individuals who trust one another and share a sense of professional purpose (Vaessen et al., 2014). Social network research also underscores the importance of network structure. Educators frequently rely on "weak ties," which are connections beyond their immediate colleagues, to access new ideas and perspectives (Hanraets et al., 2011). At the same time, network analyses of online teacher communities show that a small number of highly connected participants often function as hubs, enabling information to circulate across otherwise fragmented groups. Without these connective actors, networks tend to remain smaller and more insular, limiting the spread of innovation (Du et al., 2023).

In a descriptive study of educators in networked learning communities in the Netherlands, observations further highlighted how informal, bottom-up learning environments support the development of practice in ways that formal workshops often cannot (Hanraets et al., 2011). Networked learning is situated within teachers' everyday work, sustained through shared professional norms, and shaped by collective reflection on experience. Studies suggest that when teachers experience greater autonomy within professional networks, they are more likely to share practices, reflect on their work, and commit to sustained improvement (Vaessen et al., 2014). Trust emerges as a central mechanism: educators are more willing to experiment and adapt new approaches when learning occurs within relationships characterized by mutual respect and credibility. Over time, these dynamics can shape professional norms and support the diffusion of innovative practices.

Technology has significantly expanded the reach and flexibility of informal teacher networks, a trend that accelerated sharply during and after the COVID-19 pandemic. Online networks lower barriers related to time, geography, and institutional affiliation, making professional learning more accessible to teachers with heavy workloads. Social media platforms, online communities, and messaging tools enable rapid resource sharing and both asynchronous and synchronous peer-to-peer support at scale. At the same time, digital networks introduce new challenges, including uneven participation and the risk that learning remains invisible to formal systems. Because networked learning transcends institutional boundaries and operates outside official accountability structures, it is frequently undervalued or overlooked by school systems.

This literature suggests that informal teacher networks function as critical spaces for professional learning, particularly in complex and evolving domains such as climate education.

4.2 Examples of Informal Networks in Climate Education

Conceptually, the various informal networks examined in this chapter can be considered variants of professional learning communities (PLCs), which are groups of educators who meet regularly, share expertise, reflect, and collaborate to address teaching and learning challenges and opportunities (Serviss, 2022). In this section, we will share examples of five types of PLCs that vary in scale, structure, geographic scope, relationship to systems, and consequently in what type of professional capacity the PLC builds: school or district-based PLCs, networks of independent and faith-based schools, international & regional networks, digital and online communities, and research & educator partnerships.

4.2.1 School or District-Based Professional Learning Communities (PLCs)

School- or district-based professional learning communities (PLCs) refer to sustained groups of educators situated within a single school or district who engage in ongoing meetings to improve teaching and learning, specifically in climate change. Unlike broader networks, the geographic scope is focused on a school or district, therefore operating as contextual and implementation-focused learning spaces. They are often semi-formal and have a relationship to the formal systems within a district or school priority, helping educators translate broad goals, such as climate literacy, into deep practice changes (although examples do exist that are teacher-initiated and voluntary). Administrators or government officials may mandate a climate education goal, provide funding, or formally recognize the PLC as part of their climate education strategy. However, these networks still have elements of informal networking as participants ultimately control the agenda and adapt the goal based on their lived experiences.

In some cases, these climate-focused PLCs emerge from national policy mandates. For example, as discussed in Chap. 3, *Keep it Cool 2.0* in South Africa is a professional learning initiative aligned with the priority on mainstreaming Education for Sustainable Development into the national curriculum set by the Department of Basic Education. Between 2019 and 2022, *Keep it Cool* led the development of an online platform for resources and courses for educators that participated in an online course and further supported by a PLC. For *Keep it Cool 2.0* set between March 2024 and December 2026, the strategy is focused on upskilling district officials to facilitate professional development and setting up PLCs in schools with the self-study resources, highlighting the importance of school- and district-based PLCs to scale their initial pilot (Schudel et al., 2021; VVOB, 2024).

A second variation of governance-embedded PLCs appears at the district level. The Los Angeles Unified School District (LAUSD), one of the largest school districts in the United States, serving over 500,000 students, adopted a resolution that committed the district to integrating climate literacy and justice into all grade levels and subjects (Los Angeles Unified School District, 2025). As a result, they created and funded a Climate Literacy Champion position in each school; currently, 310 of around 1500 schools have elected a teacher as a Climate Literacy Champion. The Climate Literacy Champions bring together teachers across schools and grade levels to build climate content knowledge, share instructional strategies, and develop leadership capacity related to climate education (The Climate Reality Project: Los Angeles Chapter, 2022).

In many systems, climate-focused PLCs emerge through administrative championing rather than explicit requirements, occupying a semi-formal space within district structures. These PLCs may receive logistical support, recognition, or funding from district leaders, even in the absence of a binding policy directive. The New York City Public Schools (NYCPS) Climate Education Leadership Team (CELT) exemplifies this model, a district that serves over 900,000 students (New York City Public Schools, n.d.). Created in August 2020 by the Office of Energy & Sustainability, CELT is supported and legitimized by district leadership, yet participation and agenda-setting are driven largely by educators themselves. Structurally, CELT consists of 30 educators across NYCPS and across subjects and grade levels missioned to create resources for educators across the district, identify curricular alignment with New York State Standards, and conduct or plan professional development for the district. Nominated and voted educators chair the CELT. Rather than functioning as a compliance mechanism, CELT operates as a protected professional learning space in which teachers collaboratively build climate literacy, experiment with instructional approaches, and develop leadership capacity across the system. They also design and lead citywide climate-focused professional opportunities with more than 500 teacher attendees throughout the district while expanding partnerships with community-based organizations. Each member receives payment for their time with funding secured by the Office of Energy & Sustainability to professionalize climate education and uplift teacher action (NYC Department of Education Office of Energy & Sustainability, n.d.).

These examples suggest that school- and district-based climate PLCs operate as hybrid spaces that are legitimized by governance structures but driven by educator agency. This juxtaposition is not accidental, but essential to scaling climate education mandates. Across national, district-mandated, and administratively championed models, formal policies create direction, yet they are dependent on the informal, collaborative work of PLCs. Rather than serving as alternatives to formal reform, these PLCs appear to be a necessary mechanism through which climate education policies can move to durable practice at scale.

4.2.2 Non-governmental Networks of Independent and Faith-Based Schools

Beyond public systems and district-level initiatives, a number of non-governmental networks link independent and religious schools around common commitments to sustainability and climate action. These alliances often operate as large, distributed PLCs, providing peer-to-peer support and shared tools, similar to digital and online communities, which will be elaborated later in this chapter. They also illustrate Cohen and Mehta (2017) argument that durable reform depends on networks that build capacity and coherence across institutions rather than on isolated schools.

The Green Schools Alliance (GSA) is a peer-to-peer nonprofit "created by schools, for schools," whose mission is to connect and empower K–12 schools to lead "the transformation to a sustainable, regenerative, and climate-resilient future". It originated in 2007 when a coalition of 46 independent schools in New York City joined together to meet Mayor Bloomberg's challenge to reduce their carbon emissions by 30% by 2050, and then formalized this shared commitment as a broader alliance. Over time, GSA has expanded into a global network that has reached nearly 10,000 schools, including many independent schools and 27 of the largest public school districts in the United States that have signed its Sustainability Leadership Commitment (Green Schools Alliance, n.d.).

GSA supports climate-related school change through several mechanisms that function as non-governmental infrastructure for climate education. These include: (a) the Leadership Commitment framework that publicly commits school heads to reduce climate and ecological impact, educate and engage their communities, and transform school culture; (b) the District Collaborative, in which sustainability officers from large districts and independent schools share practices and use their aggregated purchasing power to accelerate adoption of low-carbon technologies; and (c) tools such as the START (Sustainability Tracking, Analytics & Roadmap Tool) dashboard, which helps schools benchmark emissions and other indicators and then plan and monitor whole-school sustainability efforts. In addition, the Student Conservation Community & Congress (Sc3) convenes student leaders for an intensive residential program on climate and conservation, further reinforcing youth agency across the network (Green Schools Alliance, n.d.).

Eco-Schools, run globally by the Foundation for Environmental Education (FEE), is the world's largest sustainable schools program and one of the most significant non-governmental infrastructures for climate and ESD in both public and private institutions (Foundation for Environmental Education [FEE], 2023). As of 2025 it is implemented in 101 countries, with 52,552 registered Eco-Schools and approximately 13.7 million participating students. The program is operated through national NGOs (FEE "member organizations") and an international-schools track that allows independent and faith-based schools to join even in countries without a national operator (Eco-Schools, n.d.).

Eco-Schools uses a seven-step whole-school methodology that begins with a student eco-committee and an environmental review, moves through action planning and curriculum integration, and culminates in external assessment and the award of the Green Flag. Climate change is a cross-cutting focus: FEE describes Eco-Schools as a "global sustainable schools program" that "engages the next generation in action-based learning," and has recently developed specific resources on climate resilience and school safety, providing handbooks and tools to help schools adapt to climate risks and strengthen disaster preparedness (Eco-Schools, n.d.). New Eco-Schools handbooks and online courses, developed with corporate and philanthropic partners, explicitly aim to expand "sustainability and climate change education" to tens of thousands of schools and millions of students worldwide (PPG, 2025). For independent and religious schools, Eco-Schools offers an NGO-mediated way to adopt a whole-school climate education framework without waiting for state curriculum reform, while still aligning with UNESCO's Greening Education Partnership targets (FEE, 2023).

Round Square is a global network of roughly 200 schools in about 50 countries that share an educational philosophy rooted in Kurt Hahn's experiential and character-focused approach. Member schools, many of them private or independent, such as Dunn School in California, Amman Baccalaureate School in Jordan, and Westfield School in England, explicitly present Round Square affiliation as part of their identity as fee-paying independent schools (Lower Canada College, n.d.; Stanford Lake College, n.d.).

Round Square schools commit to six IDEALS: Internationalism, Democracy, Environmentalism, Adventure, Leadership and Service. Environmentalism is defined as helping students "understand mankind's place in the universe, the forces that shape our surroundings, and the impact we have," and is operationalized through environmental and sustainability projects (Round Square, n.d.-a). The Round Square

IDEALS Challenge includes a specific Environmentalism module in which students investigate the links between behavior, global challenges and "the earth's climate," analyze environmental data, and create personal sustainability action plans, explicitly framed as responses to global environmental and climate crises (Round Square, n.d.-a).

Recent Round Square communications highlight student-led sustainability innovations across the network, such as biodiversity restoration projects on independent-school campuses, student-organized conferences on sustainable innovation, and technology-based solutions to environmental issues (e.g., using food waste to generate methane or AI to improve water efficiency) (Round Square, 2025). In these ways, Round Square functions as a non-governmental network that diffuses climate-related curricular and co-curricular practices across largely private schools via exchanges, conferences and shared frameworks rather than through state policy.

Other than networks of private schools, catholic and other faith-based school networks have also fostered climate change education professional development. The Laudato Si' Action Platform (LSAP), launched by Pope Francis in 2021 and coordinated by the Vatican's Dicastery for Promoting Integral Human Development, is a seven-year initiative that invites Catholic institutions, including schools and universities, to develop "concrete actions to protect our common home" (Mares, 2021). The platform identifies seven sectors (families; parishes/dioceses; schools; hospitals; businesses; organizations; religious orders) and seven goals, one of which is ecological education.

In the United States, the National Catholic Educational Association (NCEA), which represents Catholic elementary and secondary schools, plays a central role in the LSAP's Ecological Education working group and explicitly encourages Catholic schools to enroll, develop Laudato Si' plans, and integrate "integral ecology" into both curriculum and operations (NCEA, 2024a). NCEA positions itself as "the United States representative of Laudato Si'," curating climate-related curricula and toolkits (e.g., Wholemakers, a high-school curriculum on Catholic social teaching and climate action) and providing webinars on "Laudato Si' and Climate Action at Your High School" and on developing school Laudato Si' Action Plans (NCEA, 2024b).

At the diocesan and school level, LSAP has become a framework for multi-school climate initiatives. For example, two Catholic high schools in the Archdiocese of Newark, New Jersey, have worked with Marist Brothers ecology experts to design environmental action plans explicitly aligned with LSAP goals, supported by grant funding from their religious congregation (Catholic Schools in the Archdiocese of Newark, 2023). Similar diocesan efforts in places such as San Francisco and San Diego encourage parishes and schools to enroll in LSAP and develop multi-year climate and sustainability plans (Archdiocese of San Francisco, n.d.). Collectively, these efforts amount to a transnational, faith-based network for climate and ecological education across Catholic schools, operating largely through religious and professional associations rather than ministries of education.

Within the broader Catholic ecosystem, Jesuit networks have developed more targeted climate initiatives. The Ignatian Carbon Challenge, created by teachers at Cheverus High School in Maine and scaled by the Ignatian Solidarity Network in partnership with the Jesuit Schools Network and Catholic Climate Covenant, invites Jesuit and other Catholic high schools to "bring Laudato Si' to life" through monthly challenges for both individuals and institutions (Roewe, 2017).

Since its 2016–2017 launch, the program has engaged thousands of students across at least 26 high schools, with actions ranging from trash audits and single-use-plastic fasts to learning about the Green Climate Fund and contacting elected officials. The challenges explicitly address sustainability and climate change, encourage cross-school social media sharing (e.g., via the #iggycarbon hashtag), and include an institutional track that prompts schools to integrate sustainability into facilities, strategic planning and community formation (Roewe, 2017). In Cohen and Mehta (2017) terms, this constitutes a non-state network that creates common instructional and cultural expectations around climate responsibility across a set of semi-autonomous religious schools.

Similarly, the Don Bosco Green Alliance is a global platform for ecological initiatives of the Salesians of Don Bosco, a Catholic religious congregation with an extensive network of schools and youth centers. Launched in 2018, it is described as "an active network where young people from Salesian institutions and organizations exchange ideas and work together to address the environmental challenges our world is facing," explicitly inspired by Pope Francis's Laudato Si' (Don Bosco Green Alliance, n.d.-a). Membership is open to all Salesian institutions worldwide, and the Alliance identifies four priority dimensions, eco-spirituality, eco-education, eco-praxis and eco-justice, and three thematic priorities: combating pollution, reducing global warming, and eliminating disposable plastics (Salesian Missions, 2022).

Salesian Missions, the congregation's U.S. development arm, notes that Alliance members undertake activities such as environmental education, tree-planting and increasing green space on school properties, promoting organic agriculture and school gardens, conserving water, and expanding the use of renewable energy (Salesian Missions, 2022). A recent global campaign, #DBSchoolsGoGreen, seeks to embed these practices systematically across Salesian schools and youth works, framing ecological conversion as an integral part of Salesian pedagogy and youth ministry rather than as an add-on activity (Don Bosco Green Alliance, n.d.-b).

Together, these examples illustrate how private and religious school networks can function as powerful non-governmental intermediaries for climate education, translating theological commitments into shared frameworks, professional learning opportunities and concrete student action across thousands of schools.

4.2.3 International and Regional Teacher Networks

Beyond school- and district-based PLCs, and beyond the institutional and school networks just described, there are also cross-regional and cross-national teacher networks. These PLCs are less directly tied to formal mandates, yet are often anchored by intermediary organizations that provide legitimacy, convening infrastructure, and shared frameworks, such as intergovernmental bodies, NGOs, or other existing global teacher networks.

One form of international teacher network emerges around shared curricular frameworks, where common standards and pedagogical commitments create conditions for an international teacher network. The International Baccalaureate (IB) exemplifies this model. Environmental and sustainability are embedded across IB programmes such as their Primary Year Programme (PYP) and Middle Years Programme (MYP) and most explicitly through the "Environmental Systems and Societies" course in the Diploma Programme (secondary school). Teachers participate in IB-sponsored professional development workshops and global conferences, which establish shared language and teaching practices centered around the course's learning objectives, curriculum, and assessment (International Baccalaureate Organization, n.d.). While IB does not formally designate these interactions as PLCs, the existence of the standardized curriculum and certification process enables informal, yet persistent communities of practice to exist, including teacher-led online groups and resource-sharing networks that operate across borders. Oftentimes, these teacher networks emerge after the IB-sponsored professional development workshops. As a result, IB functions as a structuring backbone rather than a centrally managed PLC, exemplifying how PLCs can arise organically when teachers share both a curriculum and institutional identity.

Another form of international PLC is organized around network professional movements when climate education is embedded in an organization's broader commitments. Teach For All is a global network of 63+ independent national organizations with a shared mission to recruit and develop professionals into working with others to achieve systemic change, starting with a two-year commitment in high-need classrooms. Embedded in the vision "to support all children to shape a better future for themselves and others", their mission is to "expand educational opportunity around the world by increasing and accelerating the impact of social enterprises that are cultivating leadership necessary for change" (Teach For All, n.d.-a). As a result, Teach For All facilitates professional learning through cross-national communities, peer mentoring, leadership development, and shared resources. Its "Climate Education and Leadership" initiative frames climate leadership as a system-wide priority within the broader mission, thereby aiming to cultivate a critical mass of educational leaders who advance climate education across the system (Teach For All, n.d.-b). Therefore, learning occurs through a curated global resource hub, different community spaces for learning and experience-sharing, and the exchange of practices across national affiliates, often with a strong attention to local contextualization. Their global community consists of around 1500 climate education leaders that meet for monthly calls and have a WhatsApp group for exchanging updates in addition to a monthly community email that shares stories, resources, and opportunities in climate education. Workshops are also run based on community interest. Operating as a distributed, multi-learning network, this model expands the notion of PLCs beyond schools and districts, showing how transnational peer learning can support climate education even in contexts where local policy support is weak.

At a broader scale, some climate-related PLCs are convened by intergovernmental organizations, where teachers are one of several stakeholders engaged in collective learning and system change. UNESCO's Greening Education Partnership (GEP) brings together governments, civil society, youth organizations, researchers, and educators across a shared agenda to green schools, curricula, teacher education, and communities (UNESCO, n.d.-a). UNESCO's role is primarily meta-coordinative since it enables knowledge exchange across systems rather than facilitating day-to-day collaborative inquiry among teachers. Nevertheless, such platforms are important for legitimizing climate education as a global priority.

At the teacher-level, UNESCO also coordinates the Associated Schools Network (ASPnet). UNESCO's Associated Schools Network (ASPnet) Connect provides a global digital workspace for 10,000 schools in 167 countries. ASPnet schools focus on intercultural learning, peace and human rights, education for sustainable development, and other United Nations priorities. The platform offers an online directory to see what other member schools are doing, as well as a workspace where educators can collaborate, share practices, and connect with national coordinators, illustrating how online platforms can support PLC-like functions for teachers across the globe (UNESCO, n.d.-b).

Between global networks and local districts, national and regional PLCs leverage geographic proximity to support deeper collaboration among educators facing similar contexts. In the United Kingdom, the Teacher Education Network for Education for Sustainable Development and Global Citizenship (ESD/GC) was established to embed ESD/GC within initial teacher education. The network functions as a community of practice through regional groups, annual conferences, a steering committee, and shared dissemination of research and case studies. Over time, it has contributed to institutional change within teacher education programs while maintaining a strong peer-learning orientation (Inman et al., 2010). Similarly, Climate Generation's Teach Climate Network in North America combines national virtual convenings with regional cohort leadership. Regional partners lead cohort-based learning to support localized climate education efforts during their Summer Institute, helping fuel year-round collaboration of the Teach Climate Network. The Summer Institute includes keynotes, workshops, and regional cohort-based sessions led by local leaders so that educators can hone in on place-based impacts, solutions, and ongoing support needs in their communities (Climate Generation, n.d.).

These examples show that international and regional teacher networks, whether anchored by policy or broader climate education movements, connect educators across borders and contexts. At the same time, by supporting regional networks, these networks also enable meaningful place-based learning, ensuring that global collaboration informs locally relevant climate education practices.

4.2.4 Digital and Online Networks

While the previously described PLCs increasingly rely on digital tools, a distinct category of PLCs operates as fully online, geographically dispersed networks. These networks are typically untethered from formal mandates, do not rely on regional proximity, and are often sustained through voluntary participation. The defining feature is not scale or authority, but low barriers to entry that create spaces for professional learning that could otherwise be inaccessible to educators.

Some online PLCs function as field-building PLCs, connecting educators, informal educators, scientists, policy actors, and other community members together to advance climate literacy. The Climate Literacy and Energy Awareness Network (CLEAN) was supported by several US agencies and curated rigorously reviewed climate and energy education resources. At the same time, they also convene a voluntary community of practice through email lists, video calls, and an annual virtual summit called Accelerating Climate Capacity, Engagement, and Leadership Summit (ACCELS) where working groups collaborate to advance programming. Since 2008, this community has grown to 800 members from K-12 education, higher education, informal learning, workforce development, policy implementation, and other fields (Climate Literacy and Energy Awareness Network [CLEAN], n.d.). In a study of their own network, they found that there was network cohesiveness and collaboration in their climate education dimension, where the least connected dimension was climate policy (Boyd et al., 2025). What distinguishes their climate education network is that it emphasized professional learning through informal knowledge exchange including from fields outside of formal education systems.

A second form of digital PLC emerges from shared moments of disruption where educators seek connection and collective action in the absence of adequate or timely institutional support. The Climate Change Education Network (CCEN) in Australia originated during the 2019–2020 bushfires and continued during the COVID-19 pandemic. CCEN formed an open, online collective of academics and educators with regular virtual meetings, collaborative projects, shared research, and a public-facing dissemination of their work (including podcasts and curated resource libraries) (Climate Change Education Network, n.d.). CCEN demonstrated how digital PLCs can extend beyond internal knowledge exchange to knowledge mobilization, translating collective inquiry into public engagement and broader professional influence.

Digital PLCs can also kickstart from structured online professional development experiences where sustained interaction among participants is intentionally designed. Online climate change professional development programs, such as those offered by the US National Center for Atmospheric Research (NCAR) shows that virtual courses can generate robust learning communities when there is an element of inquiry-based learning and collaborative problem-solving embedded. Evaluations of these programs highlight the importance of online forums as venues for educators to share classroom experiences and co-develop instructional strategies (Johnson et al.,

2008). Similarly, in Latin America, international online initiatives such as RedOnline-IPGH demonstrate that transnational online PLCs are possible even among teachers working across multiple countries. RedOnline-IPGH was an 12-week online professional development course for in-service high school geography and social studies teachers across Central and South America. Climate change was the core topic of the online course, and the online forum board and chat room platform cultivated a community of practice. Analysis of the networking process showed how the interactions evolved from interactions to powerful collaborations even as each participant was working on individual climate change teaching plans for their own classrooms (Muñiz Solari, 2020). Overall, it is clear that online networks are a strong complement to distance learning.

Finally, some digital networks function less as conversational communities and more as platform-based infrastructure that enable professional learning through visibility and indirect peer exchange. SubjectToClimate's state hubs and online platform exemplifies this model. Rather than hosting continuous online dialogue, the platform created a shared space where educators can access vetted resources, examine exemplar lesson plans created by other educators across all grade levels and subjects, and see how climate education is taught in classrooms across contexts. In several states, there are customized state hubs specifically aligned to their state standards and contexts. These state hubs are complemented by state leads (current or former educators) and a committee of educators, higher education institutions, nonprofit organizations, and other community members, creating a pathway for platform-based networks to evolve into more interactive regional PLCs (SubjectToClimate, n.d.).

These examples suggest that digital PLCs are not merely substitutes for professional learning, but critical complements, particularly by lowering the barrier to access, enabling cross-contextual insights & collaboration, sustaining educator agency, and cultivating professional solidarity.

4.2.5 *Researcher-Educator Partnerships*

Researcher-educator partnerships are a distinct form of professional learning where teachers and researchers engage in sustained collaboration to leverage their respective strengths in instructional practice and disciplinary knowledge. Unlike other PLCs, these partnerships are typically anchored in universities and research institutes, providing legitimacy and resources. At the same time, their effectiveness relies on peer exchange and networks of trust. These partnerships function as knowledge exchange ecosystems: teachers gain access to current climate science while researchers and scientists benefit from classroom-tested insights and pathways to disseminate knowledge at scale. This reciprocity creates strong motivation for participation on both sides, providing incentives for PLCs to endure beyond formal programming.

The New Jersey Climate Change Learning Collaborative was established to implement statewide climate education initiatives, illustrating how universities can serve as a convening hub for PLCs. When New Jersey updated their student learning standards to integrate climate change across all grade levels and subjects, interdisciplinary teams based in universities partnered with nonprofit organizations to support educators in professional development. However, instead of just relying on professional development, their design intentionally includes events to convene educators, online collaboration sites for virtual networking and resource sharing, and a culminating conference, thereby creating opportunities for peer-to-peer learning and linking educators with researchers beyond technical assistance (New Jersey Department of Education, n.d.).

The Center for Sustainable Futures at Teachers College at Columbia University partners with the New York City Public Schools with similar programming. Through their Summer Institutes, Columbia University brings in scientists and researchers to conduct professional development for educators, who then leverage up-to-date knowledge to construct exemplar lesson plans for their specific discipline and grade level (Columbia LEAP, n.d.). By convening educators and researchers in a shared physical space, these programs seed professional relationships that continue informally through shared resources and ongoing collaboration between educators and researchers across the city.

Some researcher-educator partnerships are explicitly organized around the dissemination of curriculum, illustrating how professional networks function as vehicles for scaling high-quality research-based climate education materials. The Carbon TIME project was an initiative where teachers participated in multi-year networks aligned with their curriculum on carbon cycling and energy transformations. Rather than purely focusing on teacher professional development, a core part of their programming was forming PLCs through research-practice partnerships from Michigan State University so that the curriculum could adapt based on educator feedback and educators could seek support to adapt the curriculum to their local contexts (Penuel et al., 2018).

While universities often provide the institutional backbone for researcher-educator partnerships, they are not the sole prerequisite for such collaboration or network formation. Some PLCs emerge more informally. Networks such as the Polar Educators International (PEI) bring together educators and researchers from around the world in a volunteer-led global network focused on translating polar science into meaningful educational experiences. Between 2020 and 2022, the membership grew to more than 11,000 adults (Polar Educators International, n.d.). The incentive for educators to gain access to current polar research and for researchers to develop pedagogical strategies to engage diverse publics provide shared purpose and mutual benefit even in the absence of a convening body such as universities.

Altogether, researcher-educator partnerships can be sustained by PLCs to enable educators to integrate up-to-date knowledge in their pedagogy. By doing so, these PLCs help disseminate climate change education at scale in a field that is rapidly evolving and dependent on current knowledge. This becomes especially important to strengthen educator confidence in navigating the complexity of climate science and addressing persistent misconceptions.

4.3 Informal Networks as the Bridge Between Niche and Systemic Reform

The examples in this chapter illustrate how informal teacher networks operate as critical infrastructure to support teachers and schools for climate education, even in the absence of education policy support. As Cohen and Mehta (2017) note, niche reforms can operate independently from broader systemic reform, but still play a vital role in shaping it. As they argue, "it may be possible to develop such infrastructure in a niche and then gradually use it to improve the broader public system or use such things as models for similar efforts" (Cohen & Mehta, 2017, p. 27).

The cases in this chapter also suggest that the relationship between niche and system-wide reform is not sequential. Informal networks don't always precede system reform nor do they become obsolete once formal structure emerges. Instead, they reinforce one another. School- and district-based PLCs translate formal climate education policies into instructional practice or incentivize these systems to formalize climate education policy. Non-governmental networks of independent and faith-based schools, international, regional, and digital networks sustain professional learning where policy support is weak or uneven, allowing cross-national learning. Researcher-educator partnerships ensure that evolving climate science and research-based curricula can circulate beyond isolated classrooms. In every case, informal networks provide relational trust and adaptive capacity that formal systems alone cannot generate.

This chapter, therefore, reframes informal teacher networks not as peripheral, but as essential components of scalable and durable climate education reform. Even in systems with strong political will and infrastructure, climate education ultimately depends on educators enacting those priorities in practice. Informal networks function as the connective tissue that links policy to pedagogy and surfaces pedagogical innovation that can later be formalized and scaled.

References

Archdiocese of San Francisco. (n.d.). *Laudato Si Action Platform*. https://sfarch.org/laudato-si-action-platform/

Boyd, K. J., Busch, K. C., Gold, A. U., Ward, E. G., Niepold, F., Poppleton, K., Haas, D., Fiorile-Desranleau, G., & Morrison, D. L. (2025). Using social network analysis to assess connections within climate and energy education organizations: A case study conducted by the Climate Literacy and Energy Awareness Network (Clean). *Environmental Education Research, 31*(12), 2375–2398. https://doi.org/10.1080/13504622.2024.2408424

Catholic Schools in the Archdiocese of Newark. (2023, May 24). *Students respond to Pope Francis' call to 'care for our common home' with action plans*. https://www.catholicschoolsnj.org/news/students-respond-to-pope-francis-call-to-care-for-our-common-home-with-action-plans

Climate Change Education Australia. (n.d.). *Home*. https://climatechangeeducation.net.au/

ClimateGen. (n.d.). *Teach Climate Network*. https://climategen.org/teach-climate-network/

Climate Literacy and Energy Awareness Network (CLEAN). (n.d.). *Community.* https://cleanet.org/clean/community/index.html

Cohen, D. K., & Mehta, J. D. (2017). Why reform sometimes succeeds: Understanding the conditions that produce reforms that last. *American Educational Research Journal, 54*(4), 644–690. https://doi.org/10.3102/0002831217700078

Columbia LEAP. (n.d.). *LEAP summer institute.* https://leap.columbia.edu/education/summer-institute/

Coutinho, C. P., & Lisbôa, E. S. (2013). Social networks as spaces for informal teacher professional development: Challenges and opportunities. *International Journal of Web Based Communities, 9*(2), 199. https://doi.org/10.1504/IJWBC.2013.053244

Don Bosco Green Alliance. (n.d.-a). *About us.* https://www.donboscogreen.org/about-us

Don Bosco Green Alliance. (n.d.-b). *Don Bosco Green Alliance.* https://www.donboscogreen.org/

Du, H., Xing, W., & Zhu, G. (2023). Mining teacher informal online learning networks: Insights from massive educational chat tweets. *Journal of Educational Computing Research, 61*(1), 127–150. https://doi.org/10.1177/07356331221103764

Eco-Schools. (n.d.). *What is Eco-Schools?* Foundation for Environmental Education. Retrieved January 18, 2026, from https://www.ecoschools.global/

Foundation for Environmental Education. (2023). *Introduction for international schools: Education for sustainable development through the Foundation for Environmental Education.* https://www.fee.global/international-schools/about

Green Schools Alliance. (n.d.). *Our impact.* https://www.greenschoolsalliance.org/impact

Hanraets, I., Hulsebosch, J., & De Laat, M. (2011). Experiences of pioneers facilitating teacher networks for professional development. *Educational Media International, 48*(2), 85–99. https://doi.org/10.1080/09523987.2011.576513

Ignatian Solidarity Network. (2017, October 12). *Catholic high schools recognized for significant commitment to sustainability.* https://ignatiansolidarity.net/blog/2017/10/12/green-star-award-sustainability/

Inman, S., Mackay, S., Rogers, M., & Wade, R. (2010). Effecting change through learning networks: The experience of the UK Teacher Education Network for Education for Sustainable Development (ESD) and Global Citizenship (GC). *Journal of Teacher Education for Sustainability, 12*(2), 97–109. https://doi.org/10.2478/v10099-009-0057-7

International Baccalaureate Organization. (n.d.). *About the IB.* https://www.ibo.org/about-the-ib/

Johnson, R. M., Henderson, S., Gardiner, L., Russell, R., Ward, D., Foster, S., Meymaris, K. K., Hatheway, B., Carbone, L., & Eastburn, T. (2008). Lessons learned through our climate change professional development program for middle and high school teachers. *Physical Geography, 29*(6), 500–511. https://doi.org/10.2747/0272-3646.29.6.500

Mares, C. (2021, May 25). Pope Francis launches 7-year Laudato si' action plan. *Catholic News Agency.* https://www.catholicnewsagency.com/news/247777/pope-francis-launches-7-year-laudato-si-action-plan

Muñiz Solari, O. (2020). Red dinámica a través del aprendizaje online. Una experiencia Panamericana. *Revista Geográfica, 160*, 149–172. https://doi.org/10.35424/regeo.160.2019.750

National Catholic Educational Association. (2024a). *Enrollment in Laudato Si' Action Platform is in full swing.* https://www.ncea.org/NCEA/NCEA/How_We_Serve/News/Press_Releases/Enrollment_in_Laudato_Si__Action_Platform_Is_in_Full_Swing.aspx

National Catholic Educational Association. (2024b). *Laudato Si'.* https://www.ncea.org/NCEA/NCEA/How_We_Serve/Resources/Laudato_Si.aspx

New Jersey Department of Education. (n.d.). *Climate education collaboratives.* https://www.nj.gov/education/climate/collaboratives/

New York City Public Schools. (n.d.-a). *NYCPS data at a glance.* Retrieved January 18, 2026, from https://www.schools.nyc.gov/about-us/reports/nycps-data-at-a-glance

New York City Public Schools. (n.d.-b). *Climate Education Leadership Team (CELT). NYCPS sustainability hub.* Retrieved January 18, 2026, from https://sites.google.com/view/nycdoesustainability/climate-education-leadership-team

Penuel, W., de los Santos, E., Lin, Q., Marshall, S., Anderson, C., & Frank, K. (2018). Building networks to support effective use of science curriculum materials in the carbon time project. In *Networked by design* (pp. 196–214). Routledge. https://doi.org/10.4324/9781315122991-11

PPG. (2025, October 6). PPG celebrates Foundation for Environmental Education's launch of Eco-Schools materials. *PPG Newsroom*. https://news.ppg.com/Press-Releases/news-details/2025/PPG-celebrates-Foundation-for-Environmental-Educations-launch-of-Eco-Schools-materials/default.aspx

Polar Educator. (n.d.). *About us*. https://polareducator.org/about-us/

Ringling, J. J., Sanzo, K. L., & Scribner, J. P. (2021). Elementary school principals and informal learning: Leveraging networks. *Journal of Workplace Learning, 33*(2), 109–119. https://doi.org/10.1108/JWL-03-2020-0042

Roewe, B. (2017, April 21). Ignatian Carbon Challenge brings Laudato Si' to life in Catholic schools. *National Catholic Reporter*. https://www.ncronline.org/blogs/ignatian-carbon-challenge-brings-laudato-si-life-catholic-schools

Round Square. (n.d.-a). *The Round Square IDEALS challenge*. https://www.roundsquare.org/about-round-square/rs-ideals-and-discoveries/the-round-square-ideals-challenge/

Round Square. (2025, October 22). *Discovering a commitment to sustainability in Round Square schools*. https://www.roundsquare.org/articles/discovering-a-commitment-to-sustainability/

Salesian Missions. (2022, June 5). *World Environment Day: Salesian Missions highlights initiatives that care for and protect environment*. https://salesianmissions.org/salesian_press_relea/world-environment-day-salesian-missions-highlights-initiatives-that-care-for-and-protect-environment/

Schudel, I., Songqwaru, Z., Fox, H., & Nsubuga, Y. (2021). *Fundisa for Change Keep it Cool climate change education training, Nelson Mandela Bay Municipality: Course report* [Unpublished course report].

Serviss, J. (2022, May 13). 4 benefits of an active professional learning community. *International Society for Technology in Education (ISTE)*. https://iste.org/blog/4-benefits-of-an-active-professional-learning-community

Stanford Lake College. (n.d.). *Round Square*. https://www.slc.co.za/round-square

SubjectToClimate. (n.d.). *About us*. https://subjecttoclimate.org/about-us

Teach For All. (n.d.-a). *Our purpose*. https://teachforall.org/our-purpose

Teach For All. (n.d.-b). *Climate education and leadership*. https://teachforall.org/climate-education-and-leadership

The Climate Reality Project: Los Angeles Chapter. (2022.). *LAUSD resolution on climate literacy*. LA Climate Reality. Retrieved January 18, 2026, from https://www.laclimatereality.org/lausd-resolution

UNESCO. (n.d.-a). *Greening education*. https://www.unesco.org/en/sustainable-development/education/greening-future

UNESCO. (n.d.-b). *ASPnet*. https://www.unesco.org/en/aspnet

Vaessen, M., Van Den Beemt, A., & De Laat, M. (2014). Networked professional learning: Relating the formal and the informal. *Frontline Learning Research, 2*(2), 56–71. https://doi.org/10.14786/flr.v2i2.92

VVOB. (2024, April 29). *Keep it cool climate change education project 2.0*. VVOB—Education for Development. https://www.vvob.org/en/programmes/south-africa-keep-it-cool-climate-change-education-project-20

Open Access This chapter is licensed under the terms of the Creative Commons Attribution-NonCommercial-NoDerivatives 4.0 International License (http://creativecommons.org/licenses/by-nc-nd/4.0/), which permits any noncommercial use, sharing, distribution and reproduction in any medium or format, as long as you give appropriate credit to the original author(s) and the source, provide a link to the Creative Commons license and indicate if you modified the licensed material. You do not have permission under this license to share adapted material derived from this chapter or parts of it.

The images or other third party material in this chapter are included in the chapter's Creative Commons license, unless indicated otherwise in a credit line to the material. If material is not included in the chapter's Creative Commons license and your intended use is not permitted by statutory regulation or exceeds the permitted use, you will need to obtain permission directly from the copyright holder.

Chapter 5
Climate Change Education in Practice: How Schools Teach for a Warming World

Earlier chapters have argued that education and climate change are mutually shaping systems. Chapter 1 developed a systems perspective on the co-evolution of education and climate; Chap. 2 mapped first- and second-order effects and distinguished major narratives of climate change education (CCE), from climate literacy to decolonial and justice-oriented approaches. Further, Chaps. 3 and 4 examine how national governments and networks of educators support climate change education. This chapter extends the analysis to examining how climate change education is integrated into school and community practice asking:

- How, in practice, is climate change being taught in schools and through community-based initiatives around the world?
- How does this analysis compare with what large-scale reviews tell us about how environmental, climate and biodiversity issues are integrated into policies, curricula and teacher education?
- What can we learn from sites, schools, programs and communities that have made climate and sustainability a central priority rather than a marginal science topic?

We address these questions using three complementary sources:

- A review of research and global monitoring reports on how environmental themes, especially climate change and biodiversity, are integrated into national education policies, curricula and classrooms. Central among these is UNESCO's Learn for our planet (2021a) and related data from COP26, complemented by regional studies from South Asia, systematic reviews of teaching strategies (e.g., Adlit & Adlit, 2022), and national surveys like the National Center for Science Education (NCSE)'s Mixed Messages report on U.S. teachers.

© The Author(s) 2026

F. Reimers and M. Wang-Aghania, *Educating for a Climate Changed Future*,
https://doi.org/10.1007/978-3-032-23256-4_5

- Eleven in-depth case studies commissioned by the Monitoring and Evaluating Climate Communication and Education (MECCE) Project, which examine climate communication and education across highly diverse settings: Blue Flag beaches in India (Kurup et al., 2025); community deep canvassing in Canada (Burgess & Edmunds, 2025); dry-season gardening with rural women in Ghana (Atanga et al., 2025); a psycho-affective climate curriculum in France (Klein et al., 2025); participatory climate education for girls in South Africa (Murphy & Fuyani, 2025); a forest-edge madrasa in Indonesia (Desmiwati et al., 2025); a whole-school sustainable primary in Argentina (Sabbatini & Beech, 2025); a student-centered climate programme for high-school girls in Iran (Golshani et al., 2025); Dene stewardship camps in Canada's Northwest Territories (Karsgaard et al., 2025); a national "educational continuum" strategy in Senegal (Faye et al., 2025); and a youth-led "Teach the Teacher" programme in the UK (Finnegan et al., 2025).

- A cross-case analysis of 39 schools from all world regions, drawn mainly from the World's Best School Prizes nominations (environmental action category, 2022–2025). We have selected these cases not because these schools are typical, but because they sit at the frontier of what is currently possible.

Together, these sources allow us to connect macro-level initiatives, whether governmental or non-governmental (policies, curricula, teacher preparation) with micro-level practices (classroom, community, and land-based learning), and to see how different narratives of CCE are braided in practice.

The chapter is organized into three parts. Section 5.1 synthesizes the literature, global monitoring findings, and the 11 MECCE case studies on climate and environmental education. Section 5.2 offers concise portraits of 39 frontier schools. Section 5.3 integrates the literature, MECCE cases and school cases using the systems lens of Chap. 1 and the conceptual map of Chap. 2, and draws implications for system-level levers taken up in subsequent chapters.

5.1 Climate Change and the Environment in Education: What Global Evidence and In-Depth Cases Show

Research and monitoring point to a striking pattern: environmental and climate issues are more visible than ever in discourse, yet remain shallowly and unevenly integrated into policy, curricula and classrooms. Teacher preparation is limited, and pedagogical practice often emphasizes cognition over socio-emotional and action competences. At the same time, many teachers, schools and communities are innovating within and beyond these constraints.

The 11 MECCE case studies show how climate communication and education are being woven into beaches and forests, girls' education, Indigenous stewardship camps, teacher professional development, and national reform processes. They also highlight common threads: place-based learning, youth and women's leadership, Indigenous and local knowledge, and efforts to navigate the psychosocial dimensions of climate change.

5.1.1 How Far Have Environment, Climate and Biodiversity Entered Policy and Curriculum?

UNESCO's Learn for our planet report (2021a) provides the most systematic cross-national picture to date. It analyzes 78 official documents from 46 countries, 37 education sector plans (ESPs) and 41 national curriculum frameworks (NCFs), and triangulates this with interviews and an online survey of almost 1600 education stakeholders in 93 countries. Four findings are particularly relevant.

First, environmental themes appear widely, but usually superficially. Ninety-two percent of the documents contained at least one reference to environment-related themes, but 45% had little-to-no focus when depth was measured as occurrences per million words. Only 17% of documents showed a "moderate" level of attention. Standardized across document length, the average frequency was low even for the most common terms. Environmental ideas are present but rarely developed as substantial curricular threads.

Second, there is a strong skew toward generic "environment" and "sustainability", and away from climate change and biodiversity. Eighty-three per cent of documents mentioned "environment" at least once, and 69% mentioned "sustainability". In contrast, only 47% mentioned "climate change" and only 19% mentioned "biodiversity" at all. Where climate and biodiversity do appear, references are typically sparse and often not elaborated into clear learning outcomes. Updated data released ahead of COP26 reinforce this concern: UNESCO analysis of 100 countries shows that only 53% of national education curricula make any reference to climate change, and when the subject is mentioned, it is almost always given very low priority (UNESCO, 2021b). This confirms that, globally, climate change and biodiversity loss, the defining environmental crises of our time, remain marginal within the written curriculum. They are often subsumed under broad ESD language rather than treated as specific, urgent, knowledge- and action-rich domains.

Third, national curriculum frameworks engage more with environmental themes than sector plans, and there is marked regional variation. About 75% of all environment-related references were located in NCFs, and only 25% in ESPs. This reflects the fact that NCFs describe content and learning aims, while ESPs lay out broader strategic visions, but it also signals that environmental issues are rarely foregrounded in high-level education planning. Latin America and the Caribbean showed relatively high levels of environmental integration in curricula; Eastern and South-Eastern Asia and some other regions had much lower standardized frequencies.

Fourth, regional specificities reveal implementation gaps despite policy commitments. Mbah, Shingruf, and Molthan-Hill's (2022) analysis of South Asia shows that while countries like Sri Lanka and Bhutan have pledged to include climate components at different levels of their education systems, and India and Bangladesh are focusing on institutional capacity building, significant gaps remain. Policies tend to be top-down and aid-driven, failing to account for diverse socio-economic realities and local knowledge systems. For example, while 69% of South Asian youth report being "little" or "very worried" about climate change, many cannot explain global warming, having learned about it only abstractly in geography or science (UNICEF, 2021a, cited in Mbah et al., 2022). Mbah et al. propose a support framework anchored in critical pedagogy, place attachment theory, and local knowledge to bridge the gap between high-level policy targets and ground-level adaptation needs.

The MECCE case study from Senegal illustrates how one country is beginning to operationalize such a framework at system level. Senegal's Adapt'Action strategy aims to integrate climate change into the entire education and training system, from preschool through basic, secondary, vocational and higher education, using two organizing principles: an "educational continuum" and skills-based approaches (Faye et al., 2025). The strategy views climate as a cross-cutting theme that should spiral through subjects and levels, with increasing complexity, rather than appearing in isolated units. A detailed mapping of existing curricula found that climate change was mentioned mainly in science and social studies (e.g., "living in one's environment" geography), and largely absent from language, mathematics, arts and physical education. This diagnosis is being used to guide curriculum readjustments, teacher training, and extracurricular initiatives such as school gardens and community projects aimed at drought, flooding and coastal erosion. Senegal's case shows both the promise and difficulty of mainstreaming CCE across a national system: significant policy openings exist, but realizing an educational continuum for climate requires sustained inter-ministerial coordination, local contextualization and investment in teacher capacity (Faye et al., 2025).

Other MECCE cases highlight the role of institutional design in embedding climate into school identity. In Argentina, Primary School No. 12 in Mar Chiquita, a public school built as an off-grid "Earthship" beside a UNESCO biosphere reserve, has used its sustainable architecture and lagoon setting to anchor a whole-school approach to environmental and climate education (Sabbatini & Beech, 2025). Here, curriculum standards from the province are interpreted through project-based work on biodiversity, water and energy, while the building itself (solar, rainwater harvesting, food gardens) functions as a daily teaching tool. Yet the case also underscores tensions between standardized curricula, accountability systems and context-responsive, place-based climate education.

These global reviews and system-level cases suggest that policy frameworks are starting to name climate, and some countries are moving toward more integrated strategies, but depth, coherence and alignment with classroom realities remain limited.

5.1.2 *Teacher Preparation, Confidence and the "Mixed Messages" Problem*

The UNESCO report also probes teacher education and school-level practice. Across respondents, 36% reported no inclusion of environment-related content in either pre-service or in-service teacher education. Only about 30% indicated that environmental themes were included in both. Updated UNESCO/Education International data (2021b) reveal the consequence of this gap: fewer than 40% of teachers surveyed were confident in teaching about the severity of climate change, and only about one-third felt able to explain the effects of climate change on their specific region. Furthermore, 30% of the 58,000 teachers surveyed reported they were not familiar with suitable pedagogies for climate education.

More detailed national studies reinforce the picture. In the U.S., the NCSE's Mixed Messages study (2016) found that fewer than half of science teachers had any formal university coursework that addressed climate change. A follow-up study with 2014 and 2019 U.S. middle-school samples (Plutzer et al., 2024) showed modest increases in teachers' own climate coursework and professional development, but from a low base.

These preparation gaps translate into knowledge gaps. In NCSE's survey, fewer than half of U.S. science teachers knew that more than 80% of climate scientists attribute recent warming mainly to human activities. Many teachers conflated key mechanisms, prioritizing topics like ozone depletion over the greenhouse effect. Teachers' personal beliefs also matter; Plutzer et al. (2024) found that shifts in teacher acceptance of anthropogenic causes between 2014 and 2019 explained a sizable portion of the change in classroom messaging.

Perhaps the most consequential finding for students is the prevalence of mixed messages. Roughly 30% of U.S. science teachers in Mixed Messages emphasized that "many scientists" think recent warming is natural. Plutzer et al. (2024) show that a stable cohort, about 37% of teachers who teach climate, continued to present both anthropogenic and natural-cause narratives as valid scientific debates. This "balancing" strategy often privileges what Busch (2016) calls a "Science Discourse" (abstract, physical, debatable) over a "Social Discourse" (local, human, justice-oriented).

Several MECCE cases show bottom-up efforts to address these capacity gaps by reimagining teacher learning and roles.

In France, the Office for Climate Education developed a year-long, middle-school programme that explicitly sequences climate science, emotional expression and action. Teachers across 10 schools were supported through webinars and resources to implement active inquiry-based climate lessons, structured discussions of emotions, and a student-led project to green their schoolyard (Klein et al., 2025). Grounded theory analysis of repeated focus groups with students revealed persistent conceptual difficulties (e.g., confusion of climate change with generic "pollution"), but also showed that carefully scaffolded emotional work and authentic action projects increased motivation and a sense of agency. For teachers, the programme offered a model of how to move beyond purely cognitive aims and how to hold space for eco-anxiety without avoidance or despair.

In the United Kingdom, the Teach the Teacher programme flips the usual hierarchy: secondary students are trained to deliver climate education workshops to their teachers, addressing climate science, climate justice and youth perspectives on the crisis (Finnegan et al., 2025). Over 100 students have led sessions for more than 1000 teachers. Interviews with teachers indicate that this youth-led professional development can increase teacher confidence, surface the emotional dimensions of climate change, and make climate content feel relevant across subjects. The program's design acknowledges that many teachers lack formal climate training, and treats climate education as a shared learning journey rather than a one-way transmission.

In Iran, the Action to Combat Climate Change from High School (ACCCH) programme works within a highly exam-oriented system to create a seven-month, student-centered climate programme for 10th–11th grade girls (Golshani et al., 2025). Teachers adopt problem-, project- and inquiry-based pedagogies, guiding students through climate science, field trips, art-based expression and self-designed projects and campaigns. A participatory action research design shows how teacher facilitation, rather than directive instruction, was crucial for students' transformation from initial skepticism and overload to proactive climate leadership.

These cases illustrate that teacher capacity building for CCE can take unconventional forms, youth-led workshops, co-designed programs, and embedded coaching, and that addressing teachers' own emotions and beliefs about climate is as important as filling content gaps.

5.1.3 Where and How Environment and Climate Are Taught: Emerging Pedagogies

UNESCO's analysis and other reviews confirm that when climate change does enter curricula, it is typically located in science subjects (especially biology and geography) and treated primarily as a cognitive topic. A systematic review of teaching strategies from 2019–2021 by Adlit and Adlit (2022) shows that while science remains the dominant "home" for climate education, practice is diversifying and that approaches are increasingly transdisciplinary. Beyond traditional science classrooms, climate education appears in arts (using art for communication and meaning-making), humanities, and social studies. They identify several key strategies:

- Participatory and active learning: deliberative discussion, correction of misunderstandings, school or community projects.
- Socio-scientific and socio-cultural approaches: leveraging students' cultural activities and existing ideas to create personally resonant scientific learning.
- Affect-driven tactics: integrating emotional experience with cognitive learning to address eco-anxiety and empowerment.

- Digital and game-based learning: using games, digital fiction and interactive visualizations to model complex systems.
- Place-based education: utilizing local environments and hands-on activities to make abstract global concepts concrete.

The MECCE cases vividly illustrate these emerging pedagogies across formal, non-formal and informal settings.

5.1.3.1 Place-based and Livelihood-linked Education

In India, the Blue Flag Beaches initiative treats certified beaches as sites for climate communication and education. Beach managers are trained as educator-facilitators and co-design toolkits of hands-on activities, beach walks, "web of life" games, citizen science, plogging, storytelling, with interns, community members and NGOs (Kurup et al., 2025). Activities are explicitly mapped to climate concepts (e.g., sea-level rise, coastal erosion, ocean literacy), psychosocial dimensions (communities' anxieties about livelihoods), and national frameworks such as Education for Sustainable Development and Action for Climate Empowerment. The beach becomes a living classroom where tourists, school groups and National Cadet Corps youth learn to see coasts as vulnerable socio-ecological systems rather than mere recreation sites.

In Ghana, dry-season gardening in the rural community of Kuliyaa functions as a pedagogical site for women's climate learning (Atanga et al., 2025). Through a community-based participatory research design led by a Community Research Advisory Committee of women elders and knowledge keepers, the case shows how women's dry-season gardens are not only an adaptation strategy to drought and changing rainfall but also spaces for informal, experiential climate education. Newcomers learn about climate variability, soil, water and crops through working alongside experienced gardeners; elders teach indigenous pest management, seed selection and moisture conservation. Group work in the gardens supports psychosocial resilience, social support and shifts in gender roles, as women take on well-digging and become primary breadwinners.

In Indonesia, the rural madrasa MTs PAKIS was created by environmental activists and the local community to serve forest-edge children who otherwise would drop out or marry early (Desmiwati et al., 2025). The school embeds climate and environmental learning in agroforestry, biodiversity monitoring, coffee cultivation and waste management. Students learn by doing, reforesting slopes, documenting bird species, testing climate impacts on crops, and by producing creative outputs (posters, short films, a planned book) that communicate local climate realities. Here, mitigation, adaptation and conservation are inseparable from access to education and rural livelihoods.

5.1.3.2 Socio-Emotional and Action-oriented School-based Programs

The French middle-school programme already mentioned structures learning into three phases: understanding climate science through inquiry-based lessons; exploring emotions via activities using emotion cards and group dialogue; and planning and implementing a school-based adaptation or mitigation project (e.g., greening a heat-prone schoolyard) (Klein et al., 2025). Students' discourse reveals that while scientific understanding remains partial, the combination of emotional acknowledgement and concrete action opportunities can reduce paralyzing eco-anxiety and foster engagement.

The ACCCH programme in Iran similarly organizes a seven-month sequence that begins with climate science and field trips, then moves into student-chosen projects and public engagement campaigns (Golshani et al., 2025). Projects range from exhibitions and festivals to tree-planting and sustainable fashion conferences. Participatory analysis shows that students gradually shifted from viewing the class as a low-priority burden to experiencing themselves as climate ambassadors capable of influencing peers, families and officials, an example of student-centered climate pedagogy within a rigid, exam-driven system.

5.1.3.3 Indigenous, Land-based Learning

In Thaidene Nëné, on the East Arm of Great Slave Lake in Canada's Northwest Territories, Łutsël K'é Dene First Nation runs land-based stewardship camps that connect youth with elders, harvesters and scientists (Karsgaard et al., 2025). Though climate change is not always named as the explicit focus, participants learn to "read the land" as a changing system, observing ice conditions, caribou migration, water levels and fires, while acquiring practical skills (navigation, harvesting, dendrochronology, plant identification) and Dënesųłiné language. Young people describe camps as places of healing from intergenerational trauma and climate anxiety, where respect for the land is both value and practice. Climate learning here is holistic, braided with cultural continuance and Indigenous governance.

Across these diverse cases, three patterns stand out: learning is situated (in beaches, gardens, forests, lagoons), multi-dimensional (cognitive, socio-emotional, ethical, practical), and oriented toward action and survival, not just exam performance.

5.1.4 Integrative and Humanities-based Approaches

As Chap. 2 argued, powerful climate change education often emerges when science is braided with humanities, arts and civic learning. Siegner and Stapert's (2020)evaluation of a year-long, humanities-centered climate curriculum at Lowell School (Washington, DC) showed how climate themes can structure middle-school

English and social studies: students read climate fiction, wrote persuasive essays, and engaged with civics and economics, with measurable gains in literacy and strong climate engagement.

Several MECCE cases expand this picture of integrative, humanities-based CCE.

In France, the OCE programme does not confine climate to science; it uses narrative interviews and focus groups to trace how students' understanding, feelings and motivations evolve (Klein et al., 2025). The second phase, explicitly focused on emotions and hope, can be read as a humanities-inflected space where students make meaning of climate through talk, metaphor and comparison (e.g., likening future climate disasters to recent earthquakes they have seen in the media).

In South Africa, Global Challenges, Local Narratives works with Black girls in townships around Cape Town to explore climate change through storytelling, song, poetry and animation (Murphy & Fuyani, 2025). The project treats girls as co-researchers, using narrative inquiry and action research to surface their understandings, questions and communication needs. It situates climate within a web of economic precarity, racism, gendered violence and digital divides, and asks how girls can "know enough" to speak credibly in a global conversation that often marginalizes their voices. Here, climate education is explicitly public pedagogy, rooted in critical social theory and creative expression.

In the UK, the Teach the Teacher case uses digital storytelling as both research method and educational tool (Finnegan et al., 2025). Students produced short videos about their "moment of environmental awakening" and letters from their future selves in 2050. These stories reveal that informal sources, documentaries, social media, activism, were often more formative than formal schooling, and that students' imagined futures range from collapse and techno-inequality to transformative change in which nature is restored and education is re-designed. Reflective, speculative storytelling helped students articulate what they want from climate education (hope, honest science, justice, dialogue) and strengthened their sense of agency in leading teacher learning.

Outside of MECCE, SubjectToClimate is a nonprofit organization that aggregates resources and identifies their connections to all grade levels and subjects. In addition, they work with educators to create exemplar lesson plans on how climate change and sustainability can be integrated across all subjects, highlighting the opportunities with humanities-based subjects and illuminating how educators have been successfully experimenting with this approach throughout the United States. For example, an elementary teacher created an English Language Arts lesson plan in which students listen to a poem to construct their understanding of nature, explore their school to identify nature, and then leverage this understanding to write a poetry to illustrate their relationship with nature (SubjectToClimate, 2025). Another middle school educator in Oregon created a social studies lesson plan where students analyze their personal experiences with local waterways, learn the importance of water Indigenous People depend, protect, and have a reciprocal relationship with water, and then explain how they can ally with Indigenous People to protect their local waterways (SubjectToClimate, 2023).

Integrative cases like these demonstrate that climate education can, and perhaps must, engage with narrative, imagination and ethics, not only facts and graphs. Given that engagement in collective action can reduce eco-anxiety (Schwartz et. al., 2023), they also suggest that humanities-based CCE can be a powerful site for addressing eco-anxiety and for exploring contested futures.

5.1.5 Informal and Hybrid Models Linking Schools, Communities and Movements

Finally, a growing body of work examines hybrid models that connect schools with scientists, NGOs and social movements. Passow's (2022) description of the Earth2Class programme at Lamont-Doherty Earth Observatory illustrates one such model: teachers and students participate in workshops led by climate scientists, combining foundational research with classroom applications. This models a whole-ecosystem approach to professional development, helping teachers translate current science into teachable material.

The MECCE cases broaden our view of hybrid CCE beyond school-science partnerships to include community organizing, public spaces and intergenerational learning.

The Blue Flag Beaches initiative in India engages a web of actors, beach managers, research institutes, local fishers, interns, NCC cadets, tourists, in co-creating climate and coastal education (Kurup et al., 2025). Workshops and events (e.g., World Earth Day at Eden Beach, Electric Ray outreach with a marine research institute) turn beaches into hubs where scientific knowledge, indigenous fishing knowledge and visitor experiences intersect.

In Canada, the Deep Canvassing case shows how climate education and communication can be embedded in community organizing (Burgess & Edmunds, 2025). Trained volunteers have in-depth, values-based conversations with rural residents about climate and energy policies, helping them resolve cognitive dissonance and shift underlying beliefs. The case then tests ways to "escalate" supportive residents into higher-effort actions (e.g., attending community discussions) through direct mail and email. While not school-based, this work highlights psychosocial and relational dimensions of CCE and offers lessons for engaging adults in polarized contexts.

The Dene stewardship camps in Thaidene Nëné, already discussed, illustrate hybrid learning ecosystems that link Indigenous governments, Parks Canada, scientists and youth (Karsgaard et al., 2025). Camps serve simultaneously as education, research and governance infrastructure, preparing youth to become land guardians who can navigate both Dene and Western knowledge systems.

In South Africa, Global Challenges, Local Narratives connects grassroots climate activism (e.g., Black Girls Rising) with research institutions, creating a feedback loop where girls' experiences inform scholarship and advocacy (Murphy & Fuyani, 2025).

The Senegal strategy positions the Centre for Environmental Education and Training within the Ministry of Environment as a coordinating node between ministries, schools and civil society for climate education (Faye et al., 2025).

These hybrid models underscore that quality climate education increasingly happens in networks, across ministries, NGOs, communities and ecosystems, rather than within single institutions.

5.1.6 *Emerging Lessons*

The cases examined in this chapter, along with global reviews and regional analyses, underscore six broad lessons.

Environmental and climate themes are present but shallow in many systems. Most countries' policy and curriculum documents mention environment, but climate change and biodiversity are explicitly named far less often and rarely elaborated into coherent learning progressions (UNESCO, 2021a). Even when national strategies exist, as in Senegal, climate may be weakly integrated across subjects, requiring deliberate curriculum redesign and teacher support (Faye et al., 2025).

Teacher preparation is a major bottleneck, but new models are emerging. Large proportions of teachers receive no formal training in climate science or pedagogy and lack confidence in teaching climate or addressing student emotions (NCSE, 2016; Plutzer et al., 2024; UNESCO, 2021b). At the same time, youth-led professional development (Teach the Teacher), co-designed programs (OCE's French curriculum), and land-based mentorship (Dene camps) show that teacher learning can be embedded in relational, practice-rich contexts (Finnegan et al., 2025; Karsgaard et al., 2025; Klein et al., 2025).

The curriculum over-privileges cognition; effective CCE braids cognition, emotion and action. Policy documents stress knowledge acquisition; socio-emotional, ethical and action dimensions are much weaker. Yet the cases, from Blue Flag beaches and Ghanaian gardens to Iranian and French school programs, show that when learners are invited to feel, reflect and act, climate education becomes more meaningful and transformative (Adlit & Adlit, 2022; Atanga et al., 2025; Golshani et al., 2025; Klein et al., 2025; Kurup et al., 2025).

Teachers and communities often send mixed messages; addressing psychosocial dynamics is essential. In politicized contexts, balancing strategies can inadvertently legitimize non-scientific positions (Busch, 2016; Plutzer et al., 2024). Deep canvassing in Canada, digital storytelling in the UK and affect-focused programs in France and South Africa reveal that surfacing underlying values, fears and hopes is crucial for moving beyond paralysis and denial (Burgess & Edmunds, 2025; Finnegan et al., 2025; Klein et al., 2025; Murphy & Fuyani, 2025).

Context matters, including gender, indigeneity and livelihoods. As Mbah et al. (2022) argue for South Asia, climate policies must be adapted to local socio-economic realities and knowledge systems to bridge the gap between high-level goals and community adaptation. The MECCE cases show this concretely: Kuliyaa's women gardeners in Ghana, forest-edge farmers in Indonesia, Dene youth in Canada, Black girls in South African townships, and high-school girls in Iran all face distinct vulnerabilities and resources. Quality CCE in these contexts is inseparable from gender justice, Indigenous sovereignty, food security and decent work (Atanga et al., 2025; Desmiwati et al., 2025; Golshani et al., 2025; Karsgaard et al., 2025; Murphy & Fuyani, 2025).

Innovative approaches exist, but they risk remaining "boutique" without systemic support. The 39 frontier schools we analyze in Sect. 5.2, together with Mar Chiquita's Earthship school, MTs PAKIS, ACCCH, Blue Flag beaches and others, demonstrate rich, situated, justice-aware, action-oriented forms of climate education that braid multiple narratives and engage whole institutions or communities (Desmiwati et al., 2025; Golshani et al., 2025; Kurup et al., 2025; Sabbatini & Beech, 2025). Yet many are fragile, dependent on charismatic leaders, external grants or exceptional conditions. Without changes in standards, assessments, financing and accountability, they risk remaining islands of excellence in largely unchanged systems.

These patterns frame the significance of the 39 schools we examine next: they are, in many ways, operating beyond or against the constraints just described, and their practices resonate with, and extend, the MECCE cases examined previously.

5.2 Case Vignettes: How Schools Around the World Teach Climate Change

In this section we analyze the cases of 39 schools that were chosen as finalists, between 2022 and 2025, for the Environmental Education category in 'Best School Prize', an award given by the T4Education organization. Schools self-nominate for this award, and submissions are then evaluated by a panel of experts. The submissions provide extensive documentation on the activities of the school for the award for which they are being considered. While this is obviously not a random sample of schools, these cases illustrate the practices of schools that are intentionally educating students about climate.

5.2.1 The Americas

5.2.1.1 Escuela Secundaria Pdte. Domingo F. Sarmiento (Argentina)

This public, science-oriented secondary school has made environmental and climate education a core institutional mission, anchored in Argentina's 2021 Comprehensive Environmental Education Law. Beyond natural science courses, students take a dedicated environmental workshop that treats sustainability, justice, and democracy as intertwined. Four strands organize the work: (1) "Ecoideas" projects to reduce waste and pollution in the surrounding university corridor; (2) collaboration with an animal protection NGO linking animal welfare, biodiversity and environmental ethics; (3) participation in the university's ISO-14001 environmental management system, where students help measure and reduce the school's water and carbon footprints; and (4) a research project on phytoremediation, in which students sample local water bodies for heavy metals and test the use of aquatic plants to clean them. Climate change is thus taught through inquiry-based engagement with local pollution, ecosystems, and civic responsibility.

5.2.1.2 Buenos Aires Technical School Near "Ex Olimpo" (Argentina)

Located beside a former clandestine detention and torture center from the military dictatorship, this technical school fuses environmental education with memory and human rights. Through multi-year projects in healthy cooking, organic horticulture, hydroponics, aquaponics and "green chemistry," students learn to design low-impact food systems and alternative fuels such as bioethanol from native species. A flagship initiative develops an intensive urban agriculture and aquaponics system on the Ex Olimpo site under the banner "Sustaining Memory," literally creating life where there was death. Environmental degradation and climate disruption are taught as forms of structural violence; ecological restoration becomes a way of enacting "Nunca Más" ("Never Again") and reclaiming public space for justice.

5.2.1.3 EEEEM Professora Maria das Graças Escócio Cerqueira (Brazil)

In a low-funded public secondary school on the fringe of the Amazonian city of Itaituba, students have turned seed collection and tree planting into a vehicle for climate education, inclusion and hope. What began in 2022 as a small Arbor Day activity, students with disabilities germinating yellow ipê seeds in improvised nurseries, has grown into a multi-species restoration program. Students now collect seeds of native Amazonian trees, rescue wild seedlings, and raise them for reforestation projects with community partners, NGOs and federal environmental agencies. Specialized Educational Service (AEE) students lead many activities, positioning youth with disabilities as environmental protagonists. Workshops with nearby kindergartens, river clean-ups, and media outreach extend learning beyond the campus. In a neglected school with crumbling infrastructure, climate education is enacted through the very tangible work of growing future forests.

5.2.1.4 EEMTI Jaime Tomaz de Aquino (Brazil)

This rural full-day high school in Ceará, a semi-arid region, has embraced the state's Sustainable School Seal program and reorganized its work around four axes:

curriculum, environmental management, physical space and environmental communication. Teachers integrate topics such as plastic pollution, recycling and renewable energy into multiple subjects through field lessons and workshops. Students and staff co-manage school environmental initiatives: selective waste collection, water and energy saving campaigns, re-use of greywater and air-conditioner condensate, school gardens and composting. The campus has been progressively greened with native and fruit trees, rainwater harvesting systems and a biodigester prototype. Through posters, school radio and social media, students act as environmental communicators. Climate change appears not as a separate module but as a thread linking drought, waste, energy, health and local livelihoods.

5.2.1.5 Kopila Valley School (Nepal)

In Nepal's climate-vulnerable Karnali province, Kopila Valley is a free, NGO-run school that treats climate resilience and social equity as inseparable. Its highly sustainable campus, powered by solar energy, equipped with large-scale rainwater harvesting and a near-zero-waste system, serves as a daily teaching tool. Environmental themes run from early-years outdoor learning to secondary-level research projects on landslides, deforestation and regenerative agriculture. Student "Sustainability Ambassadors" lead tree-planting, waste upcycling and climate data collection; a "Changemakers' Club" works at the intersection of gender equity and climate justice, including menstrual health and eco-friendly waste management. An organic "farm to school" program supplies daily meals, involves students in regenerative farming, and extends climate-smart agriculture through partnerships with local women farmers. Graduates leave with concrete climate action experience and high rates of educational progression.

5.2.1.6 Innova Schools Pucallpa (Peru)

Located in the Peruvian Amazon, Innova Pucallpa uses a locally threatened species, the taricaya turtle, as a focal point for climate and biodiversity education. In the Taricalovers project, secondary students incubate and monitor turtle eggs provided by the protected areas authority, collect data on temperature and humidity, and oversee the release of hatchlings, achieving high survival rates. Younger students engage through stories, art and music about the taricaya and river ecosystems. The scientific work foregrounds how climate variability, habitat loss and human practices affect local species and hydrological cycles. Partnerships with indigenous Shipibo Konibo communities and regional agencies integrate indigenous knowledge and policy perspectives, positioning students as Amazonian stewards.

5.2.1.7 Hero School (Guatemala)

Hero School, serving an indigenous Maya Kaqchikel community in rural Guatemala, is itself a climate education artifact: its 20-building campus is constructed from more than 500 tons of waste, used tires, bottles and other materials, demonstrating low-carbon, disaster-resistant architecture. Within this built environment, students follow a Guatemalan-accredited curriculum infused with Education for Sustainable Development and the SDGs. Each grade level undertakes context-based projects: composting and soil building, tree nurseries, water filtration systems, smoke-safe stoves, rainwater tanks, composting latrines and entire off-grid homes for the community. Students design, build and install these structures with families in greatest need, directly addressing respiratory disease, water scarcity, sanitation and landslide risk. Climate change education is thus framed as green building, human rights and mutual aid.

5.2.1.8 ICAM—Instituto de Ciencias Agroindustriales y del Medio Ambiente (Colombia)

ICAM sits on a hillside in rural Ubaté that was heavily degraded by mining; over two decades the school has turned this land into a vast "open-air laboratory." Students and staff have reforested large areas, rehabilitated springs, installed wastewater treatment, created organic waste systems and introduced small solar and wind energy systems. An agro-environmental technical curriculum emphasizes "learning by doing" and research methods from an early age; students practice systematic field observation, draw on indigenous and peasant knowledge (e.g., medicinal plants) and communicate findings. The laboratory extends into thematic "schools" on water (supporting a network of rural aqueducts), air (monitoring pollution from coking plants and kilns with microsensors), solar radiation, noise and soils. Students' projects serve thousands of families and shape local policy debates about health and environment.

5.2.1.9 Institución Educativa Municipal Montessori—CAFELAB (Colombia)

This rural public school in Pitalito, Huila, has centered climate education on the severe pollution generated by coffee processing. Through CAFELAB, students investigate what happens to coffee pulp, mucilage and other by-products, quantify their impacts on nearby rivers and air, and then design circular-economy solutions. Guided by a structured research pedagogy, learners move from immersion (field visits, interviews with farmers) to formulation, validation and institutionalization of their findings. Student teams, deliberately led by girls as "scientific directors", prototype products such as ecological fabrics and bio-batteries made from coffee waste, documenting each with technical data sheets. The project positions climate mitigation (through waste reduction and energy innovations) as compatible with new rural livelihoods, and turns a small rural school into a regional research hub.

5.2.1.10 Institución Educativa Comercial de Envigado (Colombia)

In an urban, low-income area of Envigado, this large school responded to air pollution, dengue outbreaks and violence by reinventing itself around a socio-environmental research methodology, INVENTIPAZ. Two dedicated subjects ensure that about 20% of instructional time is devoted to student-led research and invention, guided by STEAM + Humanities, SDGs and national policy frameworks. Students, teachers and community members co-identify problems, smog, contaminated ravines, vector-borne disease, land mines, disability exclusion, and design technological solutions: solar-powered air purifying gardens, eco-orchards, solar repellents that have sharply reduced dengue cases, waste-shredding machines and robotic arms for recycling, biofuel systems, assistive devices and educational games. Climate change is taught through its interplay with health, conflict and inequality; young people experience themselves as inventors addressing multiple risks at once.

5.2.1.11 El Castillo Educational Institution (Colombia)

El Castillo serves a community affected by forest fires, water and air pollution, deforestation and unmanaged solid waste. It has adopted project-based learning and research "seedbeds" to move away from didactic teaching. A flagship initiative, the "Opportunity Shop," collects and repairs uniforms and supplies, resells them at symbolic prices to reduce waste and support low-income families, and invests profits in producing biodegradable cleaning products in the school lab. Students design and manage this and other circular-economy projects, learning about the climate impacts of linear consumption, the health consequences of burning and dumping waste, and alternatives that align environmental and social objectives. Families are deeply involved, helping shift community attitudes and practices.

5.2.1.12 New Myrnam School (Canada)

New Myrnam is a small rural K-12 school in Alberta that has turned climate and sustainability into a unifying theme for cross-curricular, multi-year projects. In a community linked to the oil economy, students and teachers collaborate on ambitious designs: an extended-season greenhouse without external heat, rainwater catchment and filtration, highly efficient hydroponics supplying local produce, an off-grid "eco classroom" powered by student-built wind, solar and biogas systems, solar-converted electric carts, and a net-zero tiny home built from a decommissioned school bus. Students conduct energy audits of municipal buildings and present proposals to the village council, several of which have been implemented.

Climate change is taught through authentic engineering, agriculture, and policy work, positioning a small rural school as a catalyst for local energy transition.

5.2.1.13 Lacombe Composite High School—EcoVision (Canada)

At Lacombe in Alberta, the EcoVision club has, over two decades, transformed the school into a living laboratory for climate mitigation and regenerative agriculture. After installing an early solar array (later destroyed by fire), students embraced the motto "Out of the ashes we will rise" and built a near-net-zero geodesic greenhouse, a two-acre food forest, aquaponics systems, composting, pollinator gardens, beehives and more solar generation. Environmental projects are tightly linked to courses in science, agriculture and trades; students can earn provincial certifications as greenhouse or beekeeping technicians. EcoVision members have also engaged in policy advocacy, co-authoring position papers and presenting recommendations to provincial ministers. Climate education here blends technical skill development, biodiversity conservation and youth policy engagement.

5.2.1.14 Bonuan Buquig National High School (Philippines)

In a coastal community in Dagupan City, this large public secondary school has made mangrove restoration the backbone of its climate and environmental education. Since 2010, students and staff have planted tens of thousands of mangroves along the degraded Longos River, learning through trial and error to select species and planting methods suited to local conditions. The maturing mangrove belt now provides fish habitat, carbon storage and natural flood and storm-surge protection. Classroom work in science, mathematics and social studies uses the project as a context for learning about coastal ecosystems, climate resilience and livelihoods. Participation in international coastal clean-ups and dumpsite rehabilitation extends learning to solid waste and pollution. Partnerships with government agencies and civil society have made the school a recognized leader in local climate adaptation.

5.2.1.15 Muntinlupa National High School (Philippines)

In dense, polluted Metro Manila, Muntinlupa NHS has built a flagship microalgae initiative, the Revitalized Algae Microfarm Project (ReVamp). Students design and maintain portable photobioreactors cultivating Chlorella vulgaris in classrooms and offices, learning how algae can absorb CO_2 and indoor pollutants at high rates. The reactors, constructed from recycled containers, LED lights, pumps, sensors, Arduino controllers and solar panels, serve as interdisciplinary teaching tools in biology, physics, engineering and computing. Students conduct research on optimal growth conditions and applications of algal biomass (e.g., biodiesel, food supplements, wastewater treatment), leading to patents and publications. ReVamp is explicitly framed in terms of multiple SDGs, and the school trains other educators and communities in algaculture, treating climate mitigation as a pathway to green livelihoods.

5.2.2 *Africa and the Middle East*

5.2.2.1 Gelan Number 2 Pre-primary and Primary School (Ethiopia)

Serving nearly 2500 low-income and migrant children near Addis Ababa, Gelan developed a climate-smart school agriculture program in response to the COVID-19 economic shock. The school set aside land for communal farming by unemployed mothers and guardians, using eco-friendly practices to grow vegetables incorporated into a school feeding program for 500+ vulnerable students. Children learn about soil, water conservation, crop cycles and weather variability through clubs and lessons based in the gardens. Mothers gain a new income stream by selling produce to the school, with reported 30% increases in household income and significant reductions in child malnutrition. Climate change education here is inseparable from food security and women's economic empowerment, teaching students that sustainable agriculture is both an adaptation strategy and a route to dignity.

5.2.2.2 Silibwet Secondary School (Kenya)

In drought-prone Nyandarua County, Silibwet has put large-scale tree planting at the center of its climate action and education. Through the Evergreen Environmental Movement (EEM) club, students have planted over 31,000 trees in 33 schools, aiming for 100,000 annually, and explicitly link this to Kenya's national goal of increasing forest cover. Planting days double as learning opportunities about the carbon cycle, the role of forests in water regulation and local climate, and the social dimensions of reforestation. The club also organizes clean-ups and supports vulnerable peers through food donations and planned scholarships, framing climate work within a broader ethic of care. The school aspires to formalize EEM as a foundation that can scale reforestation, renewable energy and climate education across the region.

5.2.2.3 Nachitukula School & Phalombe Teacher Training College (Malawi)

In Phalombe district, repeatedly hit by catastrophic floods and landslides, Nachitukula primary school and the local teacher training college have built a joint "Climate Action and Sixty Million Trees" program. Recognizing that deforestation and land degradation exacerbate flood risk, they aim to plant tens of millions of trees across thousands of hectares, using species and planting designs that support natural flood management. Teacher trainees learn to integrate climate content and tree-based adaptation into multiple subjects and to organize community planting and care. School-level tree clubs manage nurseries, monitor survival and engage families. Climate education is rooted in lived experience of disaster: floods that destroyed schools and displaced children become the starting point for understanding hydrology, land use and climate justice.

5.2.2.4 Rashaya Public High School (Lebanon)

After the COVID-19 pandemic and amid economic crisis, Rashaya High in rural Bekaa used climate-related projects to reconnect sedentary, screen-bound youth with land and community. Students and teachers transformed rocky, unused land around the school into terraced herb gardens and a beekeeping site, planting aromatic and medicinal species alongside pine trees. With guidance from a local beekeeper, students manage hives, research threats such as the Varroa mite and experiment with an AI-enabled device to control it. An automated irrigation system using moisture sensors models water-efficient, climate-smart agriculture. Science classes regularly take place outdoors, and elementary schools visit the garden as an "open-border classroom." A community-funded solar installation and plans for waste separation extend the school's climate-relevant infrastructure. Honey sales support students in financial need, linking environmental enterprise to solidarity.

5.2.2.5 Palestinian School with Environmental Science Club (West Bank)

In a West Bank village facing water scarcity, pollution and political constraints, a science teacher has built a democratic Environmental Science Club that makes sustainable living the core of school life. Students elect leaders, co-design annual plans and run projects in ecological farming, composting, water recycling, small-scale biogas, upcycling of tires and plastics, and bee conservation. They partner with national environmental agencies and NGOs, and produce recycled paper with a local factory. Climate change is taught as one of several intertwined threats, global warming, food and water insecurity, pollution, with strong emphasis on local solutions and self-reliance. Extensive documentation supports reflection and dissemination; the club has been recognized nationally and serves as a model for participatory, justice-oriented climate education under conditions of occupation.

5.2.3 Asia and the Pacific

5.2.3.1 The Arbor School (Dubai, United Arab Emirates)

Arbor is a high-resource international school that has made ecoliteracy its organizing principle. Building on the National Curriculum for England, it has developed a bespoke Eco literacy framework that treats climate change as a cross-cutting lens rather than a single topic. Six climate-controlled biodomes and wetland areas function as "living labs" where students study micro-climates, grow food and explore adaptation in an arid metropolis. Climate themes are woven through science, geography, humanities and mathematics, and the school's operations, biodiesel buses, partnerships with neutral-fuel providers, reef restoration projects like the Dubai Oysters initiative, are used as teaching tools. A robust instructional coaching program supports teachers to integrate sustainability, and an inclusion initiative ensures all students can participate. Parent networks extend practices into homes.

5.2.3.2 Dubai International Academy Emirates Hills (UAE)

This IB continuum school approaches climate education as student-driven culture rather than discrete programming. A long-standing Eco Club has led numerous initiatives, including a student-organized model COP28 simulation and a major e-waste campaign that collected nearly 3000 kg of discarded electronics and spurred installation of e-waste bins in city malls. All Primary Years Program units of inquiry are linked to SDGs, and cross-age projects, such as older students teaching water conservation to younger peers, reinforce a norm of peer-to-peer climate education. Experiential trips to Borneo, Sri Lanka and Kenya on conservation themes deepen understanding of climate-biodiversity linkages. Meanwhile, changes in operations, solar panels, campaigns for car-free commuting, make mitigation visible and measurable.

5.2.3.3 GEMS Legacy School (Dubai, UAE)

GLS, a large Indian-curriculum school, has embedded climate and sustainable development at system level. The principal and all staff completed a UN climate education course, ensuring climate literacy in every classroom. Curriculum is mapped explicitly to the SDGs and delivered largely through project-based learning. Signature initiatives include "Plant a Legacy," a global student-led tree-planting campaign engaging schools in dozens of countries; an e-waste program; organic gardening and composting; and Project Malawi, which supports school construction in climate-vulnerable communities. During the pandemic, GLS leveraged its digital strength to sustain and internationalize climate projects, and integrated social-emotional learning (PERMA-H) to address eco-anxiety. Parents are active collaborators, reinforcing sustainable practices at home.

5.2.3.4 Al Mamoura Academy (Abu Dhabi, UAE)

A co-educational primary and girls' secondary school, Al Mamoura has cultivated a student-led sustainability culture aligned with Aldar Education's environment strategy. A large Eco Committee and student sustainability ambassadors drive initiatives such as recycling challenges, campus environmental audits, and weekly "sustainability tips" in assemblies. Curriculum in science, humanities and food technology integrates topics like resource use, food miles and climate impacts, supported by partnerships with local farms and the Environment Agency. The campus hosts aeroponic gardens, a biodome and atmospheric water generators, providing real-world contexts for lessons on water-efficient agriculture and adaptation in desert climates. The school was the first in its group to earn the Eco-Schools Green Flag and now represents the network at major sustainability events.

5.2.3.5 Green School Bali (Indonesia)

Green School Bali was designed from inception as an experiment in education for sustainability. Its openair bamboo campus in a river valley minimizes environmental impact and immerses students in tropical ecosystems. Learning is organized around three "Frames": experiential ("learning by doing"), proficiency (rigorous literacy, numeracy and science) and thematic (interdisciplinary units on real-world issues), underpinned by the R.E.A.L. principle, relationship driven, experiential, authentic and local. Climate and sustainability are woven through gardening, renewable energy systems, waste reduction campaigns, social enterprises and community partnerships. Students are encouraged to propose and lead projects, from plastic-free initiatives to micro-hydro schemes, and to interrogate how their own mobility and consumption affect the island. Global attention has brought both resources and the need for critical reflection about privilege and "green" tourism, which the school addresses explicitly.

5.2.3.6 Sekolah Alam Pacitan (Indonesia)

Sekolah Alam Pacitan is a nature-based primary school whose entire pedagogical model is built around environmental behavior and local wisdom. Academic subjects are consistently linked to environmental projects through project-based and problem-based learning. A school-run "waste bank" trains students in sorting and quantifying waste; all community members bring sorted waste twice monthly. Organic waste is fully reused as compost or animal feed, and paper use has been sharply reduced through a digital platform. The campus is largely unpaved, with rice fields, vegetable plots, herb gardens and trees serving as everyday classrooms. Children water plants, feed animals, pick up litter and clean spaces daily. Strict bans on single-use plastics, creative reuse of materials, and extensive community outreach (clean-ups, green camps, workshops for parents) make climate-relevant practices habitual.

5.2.3.7 Ryan International School, Vasant Kunj (India)

Part of a large private network, this Delhi school has pursued climate-relevant innovation through student-led STEM projects and campus infrastructure. A standout initiative, D.A.B.A.T. (Drip Irrigation, Agriculture, Biofloc and Arduino Technology), responds to water scarcity and food price shocks by integrating rainwater harvesting, biofloc fish farming and sensor-controlled drip irrigation. Students built a prototype system that recycles nutrient-rich water from fish tanks to irrigate crops only when soil moisture falls below thresholds, controlled via Arduino and a mobile app. A hydroponics unit, biogas plant and vermicomposting site complement this work, connecting waste management, clean energy and agriculture. The project has received international recognition (e.g., in a model COP28 challenge), reinforcing the value of student problem-solving for climate-smart agriculture.

5.2.3.8 Delhi Public School Varanasi (India)

Since 2003, DPS Varanasi has integrated environmental stewardship into a broader ethos of "Service Before Self." Students have planted over 500 trees, transforming campus and surrounding areas. Recent initiatives explicitly align with the SDGs: the "SDGs2SaveTheWorld" project fosters global school partnerships and local collaboration with the Ganga Task Force; the Eco Brick Challenge mobilizes families to convert non-recyclable plastics into building blocks, diverting substantial waste from landfill; and a Miyawaki urban forest, developed with a government school, creates dense native vegetation on degraded land. Conservation campaigns around keystone species like Banyan and Neem link cultural and ecological dimensions. Climate change is taught as a matter of urban ecology, plastic pollution and river health, with students acting as "global change makers."

5.2.3.9 The Harbour School (Hong Kong SAR, China)

The Harbour School (THS) uses the ocean as its primary lens for climate and environmental education. Its three "centres of excellence", a Marine Science Centre, a makerspace ("The Foundry") and a 50-foot sailboat ("Black Dolphin"), enable regular fieldwork in Hong Kong waters. Marine science topics infuse science, social studies, mathematics and arts, linking coral resilience, pollution and fisheries to climate change. Students participate in citizen-science initiatives such as reef monitoring and biodiversity surveys, and have pioneered Hong Kong's first educational seaweed farm. The seaweed project, developed with global experts, teaches macroalgae biology, carbon sequestration, engineering and social entrepreneurship, with future plans to explore seaweed-based bioplastics and fertilizers. THS staff engage in regional marine education networks, amplifying the school's influence.

5.2.4 Europe and Central Asia

5.2.4.1 Finnish School of Kosovo (Prishtina, Kosovo)

Drawing on Finnish educational principles, this private school treats sustainability as a "way of life" across ages 5–18. Environmental studies is a core subject, but climate and ecological themes are woven through mathematics (e.g., carbon footprint calculations), language and arts (stories and films about environmental dilemmas) and science (ecosystems, renewable energy). Approximately 90% of students participate in eco-clubs that lead recycling drives, energy-saving campaigns and zero-waste challenges. The school has cut paper use by 40% via digitalization and eliminated about 95% of single-use plastics in the cafeteria. Creative communication, art installations, short films, social media campaigns, extends climate messaging into the community, and participation in CNN's "Call to Earth" initiative gives students a global platform.

5.2.4.2 Arkas Bilim ve Sanat Merkezi—Arkas BİLSEM (Turkey)

Arkas BİLSEM is a government-funded center for gifted students in Izmir that foregrounds climate and environmental education through project-based, interdisciplinary work aligned with the SDGs. Under themes such as "Ecosystem Awareness," students carry out field research and community surveys. Projects include "There is Sea and Sea… There is an Entire Ocean to Take Care Of," which builds ocean literacy and addresses marine plastic pollution; "Let's Upcycle: Capture the Future," which turns waste into artistic or functional objects; participation in the global Climate Action Project; and the development of educational games and radio programs on sustainability. An Environmental Team of student leaders conducts ecological audits and represents the center at national festivals. The work blends rigorous inquiry, creative expression and civic advocacy.

5.2.4.3 VBS Sint Paulus (Belgium)

This Catholic primary school in Kortrijk has turned its formerly concrete playground into a widely recognized "Climate Playground." Through a participatory design process, the school removed half its paving, planted over 40 trees and 150 shrubs, created diverse play and learning zones, and re-engineered water management so that nearly all rainwater from roofs and surfaces infiltrates or is reused on site. The new grounds include gardens, a greenhouse, composting stations, a "recycling park," birdhouses with cameras, tree "passports" accessed via QR codes and a weather station. Teachers use the outdoor environment across subjects, linking it to the SDGs and Catholic social teaching. The project has influenced city planning and inspired a regional network (BLES) supporting other schools to green their yards, positioning Sint Paulus as a model of child-scale climate adaptation.

5.2.4.4 Agora Portals International School (Spain)

On the island of Mallorca, Agora Portals has built a marine-focused climate program around its "Agora Posidonia Experience" and Aula del Mar (Sea Classroom). The Aula, designed with sensory and aesthetic care, houses a Posidonia oceanica aquarium supplied by a marine research institute; seagrass meadows serve as a concrete entry point into discussions of blue carbon and climate mitigation. A structured, age-differentiated curriculum ensures that all students, from ages 3 to 18, engage monthly with marine and climate issues and participate in intensive workshop weeks led by marine biologists. Beach clean-ups, ecological surveys and field trips to marine institutions deepen experiential learning. Classroom naming, decoration and an optional diving camp further reinforce an oceanic identity. The program has earned Eco-Schools and environmental awards and inspired replication within its school network.

5.2.4.5 International School of Nice (France)

The International School of Nice, part of the Globeducate group, has adopted the Eco-Schools framework and used its coastal location to anchor a strong marine sustainability focus. The "Flood the School" program takes students through an annual cycle of learning about marine ecosystems, analyzing human-driven threats and designing responses in collaboration with experts. A Marine Learning Hub acts as an interactive resource and exhibition space. Beyond ocean issues, the school has installed 145 solar panels, developed school gardens, run recycling and creative upcycling initiatives (e.g., Junk Kouture fashion from waste), and expanded upper-secondary offerings in Environmental Science and Environmental Systems and Societies. Student "Ambassadors of Tomorrow" lead projects on sustainable cities and water conservation and represent the school at climate summits, connecting classroom learning to public discourse.

5.2.4.6 EEEEK of Agios Dimitrios & ReUZ (Greece)

This special vocational school in a suburb of Athens has placed students with disabilities at the center of a circular-economy social enterprise. Through the ReUZ and ReUZ + initiatives, learners collect discarded household utensils via bins in partner schools, clean and repair them, and transform them into new decorative or functional items (lamps, planters, frames). Products are sold, with proceeds funding meals and support for economically vulnerable students. The project teaches resource extraction, embodied energy and waste through practice, while building practical skills in repair and design. Students also act as environmental educators, running creative campaigns, including puppet theater, videos and songs, in mainstream schools and the wider community. Climate change is framed through everyday consumption and solidarity.

5.2.4.7 International School of Zug and Luzern (Switzerland)

ISZL weaves climate and sustainability themes through experiential learning from early childhood onwards. Primary students learn in forest and pond environments and participate in zero-waste efforts. In middle school, service groups tackle recycling, gardening and composting. At the high-school level, student-led initiatives such as Youth Forum Switzerland convene young people from multiple schools to engage with the SDGs. A notable climate-related project pairs an aquaponics system at ISZL with one in a Ghanaian fishing village, offering a low-carbon alternative livelihood as climate and overfishing strain marine resources. Students travel to Ghana, co-design and install the system and study its social and ecological impacts. Partnerships with renewable-energy entrepreneurs and eco-engineering firms deepen the authenticity of projects. The school is working to adjust timetables and structures so complex, long-term climate projects can thrive.

5.3 Integrating Evidence and Examples: What Do These Classrooms Reveal?

We now analyze the global evidence reviewed in Sect. 5.1, the 11 MECCE case studies, and the 39 school cases of Sect. 5.2, using the systems lens developed in Chap. 1 and the multiple narratives about climate change education discussed in Chap. 2.

5.3.1 From Thin References to Deep Integration

UNESCO's Learn for our planet report shows that in most systems, the study of the environment, and especially climate change and biodiversity, appear only thinly in high-level documents, and mainly in generic cognitive terms (UNESCO, 2021a). Our cases show what deep curricular and institutional integration can look like.

Some schools and programs make climate and sustainability organizing principles rather than single units: Arbor School and Green School Bali; GEMS Legacy and Al Mamoura in the UAE; the Finnish School of Kosovo; International School of Nice; Mar Chiquita's Earthship school in Argentina; and MTs PAKIS in Indonesia. National strategies like Senegal's Adapt'Action seek to embed climate across subjects and stages (Desmiwati et al., 2025; Faye et al., 2025; Sabbatini & Beech, 2025). Schools anchor their identity in signature local problems, mangrove loss, coffee-waste pollution, mining damage, flood risk, dry-season crop failure, and build multi-year projects around them.

In systemic terms, these schools and initiatives have altered both the structure (what components exist and how they are connected) and the feedback loops (how actions lead to perceivable changes) of their local education sub-systems.

5.3.2 *Addressing the Cognition–Action Imbalance*

UNESCO's analysis highlights a strong bias toward cognitive learning in policy and curricula. Literature on effective CCE (e.g., Adlit & Adlit, 2022; Siegner & Stapert, 2020) similarly stresses that information alone rarely leads to action.

The cases we examined in this chapter respond to this imbalance by combining rigorous cognitive work (data collection, modelling, writing) with socio-emotional and behavioral experience (care work, solidarity, entrepreneurship, advocacy). Practices in Hero School's human-rights-framed building projects, Kopila Valley's gender-justice-inflected climate work, CAFELAB's girl-led research in rural Colombia, Kuliyaa's women's gardens in Ghana, Iran's ACCCH projects, and French and South African narrative programs all align with UNESCO's recommendation to integrate cognitive, socio-emotional and behavioral dimensions (Atanga et al., 2025; Golshani et al., 2025; Klein et al., 2025; Murphy & Fuyani, 2025).

5.3.3 *Moving Beyond "Science Discourse" Toward Multi-narrative Braids*

Busch's (2016) Science vs. Social Discourse distinction, and NCSE's findings on mixed messages, show that typical teaching privileges global physical framings and often treats causes as debatable. Our schools and MECCE cases, in contrast, tend to braid several of the narratives discussed in Chap. 2.

They combine Climate Literacy (robust science in New Myrnam, Muntinlupa's algae projects, Teach the Teacher workshops), Education for Climate Action (projects designed to conserve energy or restore ecosystems in Blue Flag beaches, MTs PAKIS, Mar Chiquita, dry-season gardens), and Critical/Decolonial approaches (connecting environmental degradation to histories of colonialism, patriarchy and violence in Dene camps, South African township projects, Ghanaian women's initiatives, and Indigenous-designed land-based programs) (Atanga et al., 2025; Finnegan et al., 2025; Karsgaard et al., 2025; Kurup et al., 2025; Sabbatini & Beech, 2025).

These braids show that the five narratives introduced earlier are not mutually exclusive; they can be combined in context-sensitive ways that are likely more powerful than any single narrative alone.

5.3.4 Teacher Capacity and Networks: From Bottleneck to Catalyst

UNESCO's review and national surveys identify teacher preparation as a key bottleneck. The exemplary schools and MECCE programs illustrate ways this constraint can be transformed.

Participation in national and global networks (Eco-Schools, Globeducate, Green Flag, Blue Flag, Teach the Teacher, etc.) and partnerships with research institutes (Earth2Class; university collaborations in Blue Flag beaches; OCE in France; Thaidene Nëné's scientist-Indigenous guardian partnerships) bring expertise into schools and communities (Finnegan et al., 2025; Kurup et al., 2025; Passow, 2022). Some schools and programs serve as training nodes, not only for teachers but also for youth, elders, beach managers and community leaders, partially addressing the professional development gap. Yet these arrangements remain patchy and heavily reliant on external champions.

5.3.5 Inequalities, Fragility and the Risk of "Boutique" Practice

The contrast between UNESCO's reviews of policy frameworks and our analysis of specific school- and community-based cases also reveals systemic challenges. Many innovative schools are well-resourced private institutions. Others achieve remarkable work under extreme constraint (Kuliyaa's gardeners, MTs PAKIS, Dene camps, township girls' programs, Blue Flag beaches), but face fragility due to reliance on volunteer labor, short-term grants or specific leaders (Atanga et al., 2025; Desmiwati et al., 2025; Karsgaard et al., 2025; Kurup et al., 2025; Murphy & Fuyani, 2025).

Without integration into curriculum standards and budgets, they risk remaining exceptional "boutique" islands in a largely unchanged system. As Mbah et al. (2022) and Faye et al. (2025) note, effective adaptation requires policy frameworks that are not just top-down but responsive to local realities and resources, and mechanisms that connect grassroots innovations to national reforms.

5.3.6 Connecting Back to the Book's Guiding Questions

This chapter began by asking what climate change "looks like" in classrooms and communities versus system-level policies. From the evidence and cases, we can say:

At system level, most learners still encounter climate as a peripheral, science-heavy topic. Teachers' knowledge and beliefs are uneven, and many send mixed messages (NCSE, 2016; Plutzer et al., 2024; UNESCO, 2021a, 2021b).

National strategies like Senegal's are promising but rare and nascent (Faye et al., 2025).

In practice, exemplary schools and community programs are demonstrating rich, situated, justice-aware, action-oriented forms of climate education that braid multiple narratives and engage whole institutions, communities and ecosystems, from beaches and forests to girls' clubs, teacher workshops and Indigenous camps.

The implication is that education can be a powerful lever for both adaptation and mitigation, but only if system structures (standards, assessments, teacher education, financing, infrastructure) are aligned to support such practices at scale, and if those structures are responsive to the diverse knowledge systems and lived realities of learners around the world. Without these system structures, such innovation remains unevenly distributed and fragile. Chapter 7 examines those system-level levers in detail.

References

Adlit, M. F., & Adlit, M. F. (2022). Teaching climate change: A systematic review from 2019–2021. *International Journal of Advanced Research and Publications, 5*(5), 12–19.

Atanga, R. A., Acharibasam, J. B., & Kabah, K. P. A. (2025). *Dry season gardening as climate learning by rural women*. MECCE Project Funded Case Study. https://mecce.ca/publications/dry-season-gardening

Burgess, M., & Edmunds, N. (2025). *Changing the climate and energy conversation through deep canvassing*. MECCE Project Funded Case Study. https://mecce.ca/publications/deep-canvassing

Busch, K. C. (2016). Polar bears or people? Exploring ways in which teachers frame climate change in the classroom. *International Journal of Science Education, Part B, 6*(2), 137–165. https://doi.org/10.1080/21548455.2015.1027320

Desmiwati, Asyief, K. B. M., As'attohara, B. P., Novianti, W., & Christian, Y. (2025). *Teaching young people in a forest community: Linking climate education with local realities*. MECCE Project Funded Case Study. https://mecce.ca/publications/teaching-young-people-forest-community

Faye, A., Fall, M., Sarr, M., Lo, H. M., & Faye, T. (2025). *Climate change and the educational continuum: Mainstreaming climate change education and training policy at the national level*. MECCE Project Funded Case Study. https://mecce.ca/publications/mainstreaming-climate-change-education-policy

Finnegan, W., Drayson, R., Guruprasad, S., & Runkle, Q. (2025). *Empowering young people to teach their teachers: Discovering quality climate education through digital storytelling*. MECCE Project Funded Case Study. https://mecce.ca/publications/teach-the-teacher

Golshani, Z., Masoumi, S., Chalabiyani, B., & Alavijeh, E. E. (2025). *A student-centred climate education program for high school girls in Iran*. MECCE Project Funded Case Study. https://mecce.ca/publications/climate-education-in-iran

Karsgaard, C., Michel, L. J., & Catholique, I. (2025). *Climate change and futures-building through Dene stewardship camps*. MECCE Project Funded Case Study. https://mecce.ca/publications/dene-stewardship-camps

Klein, S., Picasso, M., Pichon, C., & Casati, R. (2025). *Emotions, climate change, and the classroom: A resource for primary school teachers*. MECCE Project Funded Case Study. https://mecce.ca/publications/emotions-climate-change-and-the-classroom

Kurup, S., Joshi, M., & Menon, S. (2025). *Blue Flag beaches as sites for place-based climate communication and education.* MECCE Project Funded Case Study. https://mecce.ca/publications/blue-flag-beaches

Mbah, M. F., Shingruf, A., & Molthan-Hill, P. (2022). Policies and practices of climate change education in South Asia: Towards a support framework for an impactful climate change adaptation. *NPJ Climate Action, 1*, Article 28. https://doi.org/10.1007/s44168-022-00028-z

Murphy, M., & Fuyani, X. (2025). *Global challenges, local narratives: Participatory climate education for girls.* MECCE Project Funded Case Study. https://mecce.ca/publications/global-challenges-local-narratives

National Center for Science Education. (2016). *Mixed messages: How climate change is taught in America's public schools.* National Center for Science Education.

Passow, M. J. (2022). Earth2Class: An innovative strategy to teach climate change education. *Creative Education, 13*(3), 861–875. https://doi.org/10.4236/ce.2022.133050

Plutzer, E., Branch, G., & Townley, A. L. (2024). Climate change education in U.S. middle schools: Changes over five pivotal years. *Journal of Microbiology & Biology Education, 25*, e00015-24. https://doi.org/10.1128/jmbe.00015-24

Sabbatini, C., & Beech, J. (2025). *The whole-school approach to climate change education in action: A year with Primary School No. 12 Mar Chiquita.* MECCE Project Funded Case Study. https://mecce.ca/publications/whole-school-approach-in-action

Schwartz, S. E. O., Benoit, L., Clayton, S., Parnes, M. F., Swenson, L., & Lowe, S. R. (2023). Climate change anxiety and mental health: Environmental activism as buffer. *Current Psychology, 42*(20), 16708–16721. https://doi.org/10.1007/s12144-022-02735-6

Siegner, A., & Stapert, N. (2020). Climate change education in the humanities classroom: A case study of the Lowell School curriculum pilot. *Environmental Education Research, 26*(4), 511–531. https://doi.org/10.1080/13504622.2019.1607258

SubjectToClimate. (2023, March 26). *Indigenous peoples and waterways lesson.* SubjectToClimate. https://subjecttoclimate.org/lesson-plans/indigenous-peoples-and-waterways-lesson?queryUid=query_1769689325737_stc_materials_yvucz3gix

SubjectToClimate. (2025, March 24). *Nature and the self poetry lesson.* SubjectToClimate. https://subjecttoclimate.org/lesson-plans/nature-and-the-self-poetry-lesson?queryUid=query_1769689066946_stc_materials_s6efhh48j

UNESCO. (2021a). *Learn for our planet: A global review of how environmental issues are integrated in education.* UNESCO.

UNESCO. (2021b, October 31). *Only half of the national curricula in the world have a reference to climate change, UNESCO warns.* UNESCO. https://www.unesco.org/en/articles/only-half-national-curricula-world-have-reference-climate-change-unesco-warns

UNESCO Mahatma Gandhi Institute of Education for Peace and Sustainable Development. (2017). *Textbooks for sustainable development: A guide to embedding.* UNESCO MGIEP.

Open Access This chapter is licensed under the terms of the Creative Commons Attribution-NonCommercial-NoDerivatives 4.0 International License (http://creativecommons.org/licenses/by-nc-nd/4.0/), which permits any noncommercial use, sharing, distribution and reproduction in any medium or format, as long as you give appropriate credit to the original author(s) and the source, provide a link to the Creative Commons license and indicate if you modified the licensed material. You do not have permission under this license to share adapted material derived from this chapter or parts of it.

The images or other third party material in this chapter are included in the chapter's Creative Commons license, unless indicated otherwise in a credit line to the material. If material is not included in the chapter's Creative Commons license and your intended use is not permitted by statutory regulation or exceeds the permitted use, you will need to obtain permission directly from the copyright holder.

Chapter 6
What Do Students Around the World Know About Climate Change, and What Should Be Taught and How?

Around the world, youth are often portrayed as the generation that will lead the fight against climate change. Yet, beneath this narrative lies a crucial question: *what do students actually know about climate change?* Relatedly, what should they be taught? And how?

Large-scale cross national assessments of students' knowledge and skills such as the OECD Programme of International Student Assessment (PISA) 2018 reveal that while most students are "environmentally enthusiastic," meaning that they display awareness, self-efficacy, and a sense of purpose, there are widely varying levels of knowledge as well as involvement with environmental actions (OECD, 2022). In addition, other large-scale international studies, such as the International Civic and Citizenship Education Study (ICCS) and the Trends in International Mathematics and Science Study (TIMSS) from the International Association for the Evaluation of Educational Achievement (IEA), assess student environmental attitudes through self-reports of values, responsibility, perceived threat, and intended or reported behaviors rather than direct measures of climate knowledge or skills. Similar to PISA, these cross-national surveys reveal a gap between expressed responsibility and concrete civic engagement (Isac et al., 2025). These gaps are not coincidental; they reflect the fragmented ways in which climate change is taught, the result of the absence of the policy systemic coherence we propose in the next chapter. As earlier chapters explained, in many contexts, climate topics are confined to the sciences, limiting opportunities to explore ethical, civic, or justice-oriented dimensions. Beyond the systems, infrastructures, networks and school practices that support climate literacy (topics addressed in Chaps. 3–5 and 7) it is important to establish what students actually know and what misconceptions currently exist. Without this empirical grounding, policy conversations risk being built on assumptions about the likely effects of intentions and actions.

This chapter, therefore, examines the global state of student knowledge, attitudes, and engagement with climate change. It then explores what kinds of educational experiences improve understanding and motivation, highlighting both the challenge

© The Author(s) 2026

F. Reimers and M. Wang-Aghania, *Educating for a Climate Changed Future*,
https://doi.org/10.1007/978-3-032-23256-4_6

and progress in building climate literacy. The final section synthesizes frameworks on what and how to teach about climate change, integrating insights from the gaps identified earlier to offer a practical foundation for strengthening climate literacy.

6.1 Student Knowledge, Attitudes, and Behavior Towards Climate

The capacity to understand and respond to climate change encompasses several distinct dimensions: (1) general knowledge and skills that enables access to subject specific knowledge about climate, such as understanding of probabilities, and of complex causation; (2) subject specific knowledge about climate; (3) attitudes towards the environment and climate, and (4) dispositions to act based on one's knowledge.

The first dimension involves general knowledge and skills such as systems thinking, probability, and reasoning about complex causation. Without these, students struggle to grasp long time scales, feedback loops, or the probabilistic nature of climate projections. The second dimension, subject-specific knowledge about climate systems, impacts, and solutions, includes disciplinary and interdisciplinary knowledge: understanding the greenhouse effect, the carbon cycle, and the links between climate, health, food systems, and migration. The third dimension, attitudes toward the environment and climate, includes responsibility, concern for others, justice, and care for the planet. Climate change education, in this view, is not value-neutral: it necessarily engages questions of intergenerational equity, global inequality, and what constitutes a "good life" within planetary boundaries. Finally, the fourth dimension, dispositions to act based on one's knowledge, acknowledges that responding meaningfully to climate change requires students to create new value (for example, by imagining and testing low-carbon ways of living), reconcile tensions and dilemmas (such as those between economic development, personal convenience, and ecological limits), and taking responsibility for their own choices while engaging in collective action.

This is the kind of multidimensional framework that guided the design of the 'World Course' an interdisciplinary K-12 curriculum aligned with the UN Sustainable Development Goals (Reimers, 2020; Reimers et al., 2016) and that undergird UNESCO's Commission on the Futures of Education Report Reimagining Our Futures Together as well as the OECD's Learning Compass 2030. UNESCO's International Commission on the Futures of Education frames climate change education as requiring an integrated set of competencies that go well beyond scientific knowledge alone. In Reimagining our futures together, the Commission argues that learners need ecological and systems understanding of human interdependence with a "living, damaged planet," supported by interdisciplinary and intercultural learning, as well as scientific and critical digital literacies to discern misinformation and denial. It also emphasizes civic agency and democratic participation, capacities to deliberate and act collectively, as well as an ethic of care, solidarity, and justice-oriented reasoning (including gender-responsive and intersectional attention to unequal climate burdens) so that education can build capabilities for adaptation, mitigation, and reversal of climate change (UNESCO, 2021).

The Learning Compass argues that student agency, the capacity of learners to set goals, make informed choices, and take responsible action in order to shape their own lives and the broader world should be the core of the curriculum. Around this core, the Compass identifies interrelated domains of learning: core foundations (including literacy, numeracy, digital and data literacy, and physical and mental health); different forms of knowledge (disciplinary, interdisciplinary, epistemic and procedural); a broad repertoire of cognitive, social-emotional and practical skills; and attitudes and values oriented toward individual, societal and environmental well-being. These elements are mobilized through three "transformative competencies", creating new value, reconciling tensions and dilemmas, and taking responsibility, within an ongoing cycle of anticipation, action, and reflection (OECD, 2019). The Learning Compass also makes environmental quality and sustainability part of its destination of "well-being" for individuals, communities, and the planet, in explicit alignment with the Sustainable Development Goals. From this perspective, climate literacy is best understood as the ability to mobilize core foundations, climate-related knowledge, skills, attitudes, and values through an anticipation–action–reflection cycle: anticipating climate risks and futures, taking informed individual and collective action, and reflecting critically on outcomes to improve future decisions.

Throughout this chapter, then, the question is not only what facts students can recall about climate change, but to what extent education systems are enabling them to develop this multidimensional set of competencies needed to navigate, and to help shape, a climate-changed world.

In addition to the OECD and IEA cross-national large-scale assessments, scholarly literature reveals consistent patterns in student knowledge and misconceptions about climate change across national contexts. This chapter draws on both sources of evidence: international assessment data are used to establish a broad empirical baseline, while insights from the research literature provide finer-grained analyses of how students conceptualize climate change. The synthesis that follows is grounded primarily in four comprehensive literature reviews, selected for their methodological rigor, breadth of international coverage, and recency, though some of these draw on primary studies dating back to the 1990s. While efforts were made to include studies spanning diverse regions and grade levels, the literature remains heavily concentrated in the United States and the United Kingdom, underscoring a need for more empirical research on student climate understanding in underrepresented geographic regions:

- Bhattacharya et al. (2021)—This systematic analysis reviewed 178 empirical, peer-reviewed students in formal K-16 settings between 1991–2018, though the majority was published between 2010–2018. These studies came from 27 countries, with the majority focused on secondary students. While this study looked at trends in education research as well as teacher knowledge and beliefs about climate change, we will focus on the synthesis regarding student knowledge, conception, beliefs, and learning about climate change.

- García-Vinuesa et al. (2024)—This meta-synthesis literature review analyzes 76 quantitative results or qualitative findings published between 1993–2017 among secondary school students, analyzing their knowledge, beliefs, perceptions, and attitudes about climate change.
- Lee et al. (2020)—This narrative synthesis reviewed 51 peer-reviewed journal articles regarding youth between 6 and 18 years regarding their perceptions, beliefs, attitudes, and conceptualization of climate change causes, impacts, and solutions between 2007 and 2018. While the majority of the studies came from high income countries and predominantly USA, UK, and Australia, the studies came from over 20 countries.
- Wildbichler et al. (2025)—This systematic review of literature leverages PRISMA guidelines to analyze and synthesize 44 peer-reviewed studies between 2013–2022. We primarily look into the synthesis for research question 2 and 3: which students' ideas about the scientific underpinnings of climate change are reported in the literature and which ideas are particularly prevalent throughout the analysed literature. Most of the studies were about general understanding of climate change.

Collectively, these four reviews offer an evidence base for understanding global patterns in student knowledge, attitudes, and behaviors regarding climate change. The following synthesis draws on their overlapping findings.

6.1.1 Knowledge: Causes, Impacts, and Solutions

In general, available cross national studies reveal that student knowledge about climate change has not improved over time. According to PISA 2006 and PISA 2015, knowledge about environmental science has decreased approximately 0.5 percentage points on average across all OECD countries (OECD, 2022). This knowledge of climate change varied across countries. In fact, there was a trend that lower-income countries and those most susceptible to climate change impacts scored lower in environmental knowledge. In some countries like Canada, Hong Kong, Scotland, Singapore, Spain and Chinese Taipei, most students were able to respond correctly to environmental sustainability questions, while in others like Albania, Indonesia, Kazakhstan, Morocco, Panama, and the Philippines, only a minority could do so (OECD, 2022). Many studies have also suggested that knowledge about climate change has decreased in the past decade (Bhattacharya et al., 2021).

Complementing OECD's PISA findings, the international assessments conducted by the International Association for the Evaluation of Educational Achievement (IEA) provide another important lens on what students know about the environment and, indirectly, about climate change. In TIMSS 2019, IEA constructed an environmental knowledge scale from science items at grades 4 and 8 that deal with conservation and use of natural resources, ecosystems and interdependence, pollution and its consequences, weather and climate (including the distinction between weather and climate and evidence of climate change), and natural cycles such as the water and carbon cycles (Isac et al., 2025; Mullis et al., 2020). Analyses of this scale show substantial variation between countries: in some systems, average scores exceed 550 points, well above the international mean, while in others they do not reach 400, a full standard deviation lower (Isac et al., 2025). Within countries, gaps of more than 100 points between lower- and higher-achieving students are common, and environmental knowledge is strongly stratified by socio-economic status and urban–rural location (Isac et al., 2025). Overall, TIMSS suggests that many students have only a partial understanding of key environmental and climate-related concepts by the end of primary and lower secondary school, and that this understanding is unevenly distributed (Isac et al., 2025; Mullis et al., 2020).

PIRLS, IEA's reading literacy study, does not report a dedicated environmental or climate scale, but it underscores an important precondition for climate literacy: the ability to comprehend and critically evaluate complex informational texts (Mullis et al., 2017). In practice, the scientific and civic understandings reported through TIMSS and ICCS presuppose the kinds of informational reading proficiencies documented in PIRLS. Taken together, these IEA studies suggest a consistent picture: by the end of primary and lower secondary school many students recognize environmental and climate threats and endorse environmental protection as a civic value, but their conceptual understanding of environmental and climate systems is often shallow and highly unequal, and relatively few report sustained engagement in collective environmental action (Isac et al., 2025; Mullis et al., 2017, 2020; Schulz et al., 2018, 2025).

More specifically, according to García-Vinuesa et al. (2024), secondary school students' knowledge revolved mostly around the general causes and impacts rather than solutions. However, students tended to understand the impacts of climate change more than the causes, and in fact, carried more misconceptions about the latter (Lee et al., 2020; Wildbichler et al., 2025). Many students can identify that human activities increase carbon dioxide levels, and carbon dioxide consistently appears as the most widely recognized greenhouse gas (Lee et al., 2020). While most studies show broad agreement that human activities contribute to climate change, Bhattacharya et al. (2021) highlight a more nuanced pattern in the USA: a study of middle and high school teachers surveyed in the US believed that climate change was due to natural causes. Although this is about educators rather than students, the Aspen Institute surveyed US students in 2024 and found that only 54% of teenagers identified greenhouse gas emission from human activity as the primary contributor to climate change (This is Planet Ed, 2025). This contrasts with patterns in other countries, where anthropogenic causes were more widely accepted.

Students frequently point to visible or familiar human activities in their context such as transportation, industrial production, or energy use, as primary contributors to climate change, often reflecting what they observe in their own local environments (García-Vinuesa et al., 2024). Wildbichler et al. (2025) similarly find that students across countries commonly cite industry, factories, transport, and energy supply as main causes. While students can name these sources, they generally cannot describe the underlying sciences and causal relationships (Lee et al., 2020) as well as scientific concepts critical to understanding climate change such as atmospheric gases, energy budget, ozone layer, enhance greenhouse effect, and the carbon cycle (Bhattacharya et al., 2021).

In contrast, whereas causes are associated with human activity, students tend to recognize the physical and environmental impacts of climate change rather than the social, economic, and health-related consequences. The research reviewed consistently shows that when students are asked to describe the effects of climate change, they recall visible, media-salient impacts such as melting ice caps, rising temperature, sea-level rise, species extinction, and landscape changes more than systemic disruptions to food systems, livelihoods, health, or migration (García-Vinuesa et al., 2024; Lee et al., 2020; Wildbichler et al., 2025).

As for climate solutions, students frequently cite tree-planting, reducing vehicle emissions, promoting renewable energy, and recycling as primary climate solutions (García-Vinuesa et al., 2024; Lee et al., 2020; Wildbichler et al., 2025). These actions appear repeatedly in both open-ended and close-response formats, suggesting that students might rely on widely circulating environmental messages rather than climate-specific knowledge. For example, while many recognized planting trees as a solution, there was a lack of understanding of the exact mechanisms for which this was related to climate change and thereby, students could not explain the lag time involved in this solution (Lee et al., 2020). Moreover, recycling is often overestimated as a solution (This is Planet Ed, 2025).

According to García-Vinuesa et al. (2024), students tend to suggest solutions that other actors like governments, industries, or international organizations could undertake. This aligns with findings from Lee et al. (2020), who observe a recurring pattern in which participants propose large-scale actions for which they bear no direct responsibility besides recycling. This externalization may reflect both a realistic sense of the scale of climate mitigation and a perception that individual behavior plays a limited role, but it also highlights a pedagogical gap: students seldom articulate how individually, community, and systemic actions interact as solutions.

Overall, there is a lack of depth on scientific concepts behind climate change. As a result, there are common misconceptions that are globally widespread. Across all studies, the most common involves falsely associating a primary cause of climate change with ozone layer depletion, including the belief that "holes in the ozone layer" worsen the greenhouse effect (Bhattacharya et al., 2021; García-Vinuesa et al., 2024; Lee et al., 2020; Wildbichler et al., 2025). This confusion has persisted for decades (Bhattacharya et al., 2021).

Another misconception is the confusion between climate and weather, especially the long time scales associated with climate (García-Vinuesa et al., 2024; Wildbichler et al., 2025). This underscores how the long temporal scale and broad spatial dimensions of climate remain difficult for students to conceptualize. Other misconceptions include conflating climate change with acid rain, seismic activity, and pollution more broadly (García-Vinuesa et al., 2024; Wildbichler et al., 2025).

6.1.2 Attitude

IEA's civic education study, ICCS, adds a complementary perspective on students' environmental awareness, attitudes, and engagement. ICCS 2016 introduced "environmental sustainability in civic and citizenship education" as a specific focus area, asking grade-8 students both how they viewed environmental threats and how they understood environmental responsibility (Schulz et al., 2018). Across participating countries, a majority of students saw issues such as pollution, water and food shortages, infectious diseases, poverty, and climate change as major threats to the world's future (Schulz et al., 2018). About half (49%) regarded "making personal efforts to protect the environment" as a very important part of being a good adult citizen, yet only 9% reported having participated in an environmental action group in the previous year, an early indication of the attitude–action gap that will recur throughout this chapter (Schulz et al., 2018). Curriculum and teacher data from ICCS show that most systems now include understanding environmental issues among the stated aims of civic and citizenship education, and teachers frequently rank "promoting respect for and safeguard of the environment" among their most important goals for this learning area (Schulz et al., 2018). More recent ICCS 2022 results reinforce these patterns: large majorities of students view pollution, water shortages, climate change, and biodiversity loss as significant threats, and students with higher civic knowledge tend to express greater concern and stronger support for environmental protection (Schulz et al., 2025).

Findings from OECD's PISA assessments align with this pattern: there is a correlation between pro-environmental attitudes and PISA science test scores, implying that either pro-environmental attitudes motivate learning or learning informs pro-environmental attitudes (OECD, 2022). This is substantiated by other studies reviewed by García-Vinuesa et al. (2024): belief in climate change and anthropogenic causes of climate change increases when students are taught rigorous scientific knowledge. Further, across studies, students consistently believe that climate change is occurring with concern on its consequences, with female students feeling greater levels of concern (Bhattacharya et al., 2021; Lee et al., 2020). This concern is also stronger in lower-middle and upper-middle income countries such as Turkey, India, and China than in high-income countries like the U.S., U.K., and New Zealand. There are, however, nuances in this pattern. For example, in some studies, there is high belief in climate change as well as a moral sense of responsibility in Norway, while in Brazil, only around 30% accept this responsibility (García-Vinuesa et al., 2024). While sociopolitical factors may play a role, psychological distance and personal relevance may also play a role. As noted in the section about knowledge, students have a hard time conceptualizing the temporal and spatial vastness of climate change, making it difficult to associate its existence and risk to everyday experiences.

Lee et al. (2020) specifically note a pattern that students who held "communitarian" rather than "individualistic" worldviews also believed in climate change. While this study was done in the US, this may explain why students in different countries with a traditionally collectivist attitude tend to believe in climate change. However, when controlled for knowledge, these factors did not play a huge role, thereby noting the power of knowledge in affecting attitudes. This pattern between climate literacy and belief in climate change is also recognized by Bhattacharya et al. (2021)'s review.

In some studies, age plays a role in climate change attitudes: younger children tend to believe, worry, and take more responsibility for climate change (Lee et al., 2020). Some studies attribute this to the "adolescent dip," which is when developmental changes in moral reasoning may increase awareness of personal costs or increase the sense of powerlessness for such a large problem (Prati et al., 2022).

An interesting pattern was that students value formal climate education as both a trusted source of information and a site for action. Bhattacharya et al. (2021) report that students consistently viewed school-based learning as essential for understanding the climate system. García-Vinuesa et al. (2024)'s findings reinforce this pattern: students trust schools and educational institutions as uniquely capable of driving meaningful climate action.

6.1.3 *Behavior*

Despite expressed concern and consensus on the existence of climate change, there is a persistent attitude-action gap: PISA data show that while many students hold pro-environmental values, only about one-fifth report active involvement in environmental behaviors, a pattern the OECD describes as "environmental misalignment" (OECD, 2022).

When we analyze this environmental misalignment, both García-Vinuesa et al. (2024) and Lee et al. (2020) noted that students gravitate toward low-cost, low-disruption behaviors regardless of country. In other words, students tend to adopt actions that are convenient or require little personal sacrifices, such as unplugging appliances, recycling, or planting trees, even when students simultaneously acknowledge that these actions are not the most effective. This aligns with the broader pattern of students overemphasizing familiar, visible actions and underestimating the scientific or systemic impacts of more consequential measures.

In particular, students from higher income countries were more reluctant to adopt high-impact behaviors that entailed lifestyle changes such as reducing meat consumption, buying fewer goods, or using public transport. Cultural norms play a role here as in some societies, actions like reducing consumption may be interpreted as a marker of poverty, while in other countries, students questioned whether individual behavior had any meaningful effect on global climate outcomes.

Most of these studies, however, are published before 2020, leaving limited empirical evidence of how today's students, growing up amidst intensifying climate impacts, understand the issue.

6.2 Effective Educational Experiences

The findings from these literature reviews and large-scale assessment reveal gaps in knowledge, attitudes, and behavior. These findings raise a critical question: what kinds of learning experiences can meaningfully improve such comprehensive climate literacy?

Across diverse contexts, from elementary to undergraduate learners and short interventions to full curricula, there are consistent themes regarding effective climate change education. Research converges on a set of pedagogical principles that support knowledge gains, motivation, and, to a lesser extent, behavior change. For instance, a recent meta-analysis of 53 climate change education interventions targeting primary and secondary students found that knowledge outcomes showed the largest improvements compared to attitudes and behaviors (Aeschbach et al., 2025). Despite these gains, there is a limited understanding of the mediators that drive these outcomes.

Earlier systematic reviews provide additional insights. Monroe et al. (2019) synthesized studies prior to 2017 and identified two overarching themes common to most environmental education: "(1) focusing on personally relevant and meaningful information, and (2) using active, engaging teaching methods." They also highlighted four themes: "(1) engaging students in deliberative discussions, (2) creating opportunities to interact with scientists, (3) explicitly addressing misconceptions, and (4) implementing school- or community-based projects."

Building on these foundational reviews, this section examines more recent empirical evidence to identify common approaches that move the needle on climate knowledge, attitudes, and behaviors.

6.2.1 Focusing on Scientific Literacy and Basic Climate Knowledge

As Aeschbach et al. (2025) notes in their meta-analysis, interventions that included explicit instruction on foundational climate science, such as the difference between climate and weather, had a bigger effect on knowledge outcomes overall. This finding is consistent based on knowledge gaps students often have (as elaborated earlier in this chapter): without a conceptual grounding in climate systems, students cannot meaningfully grasp more complex ideas such as feedback loops, anthropogenic forces, or the temporal and spatial scales at which climate change operates. Therefore, it is important to understand what types of interventions are effective in providing the scientific foundations for climate change.

Understanding complexity is commonly researched in science education. Climate change education magnifies these challenges not only because the complexity of the subject, but also because public discourse around climate change is sometimes politicized and saturated with misconceptions. As noted earlier in the chapter, misunderstandings such as conflating climate change with the ozone hole is widespread.

Monroe et al. (2019) identified a pattern among studies: explicitly addressing misconceptions through a constructivist approach is effective. This involves eliciting students' prior knowledge to identify misconceptions, using discussion or hands-on activities to challenge them, and then, isolating instruction to a small number of core ideas necessary to build a more scientifically accurate conceptual model.

A recent study among 248 secondary students revealed an interesting finding in developing climate change knowledge: using novel climate data can help build climate knowledge. In this study, an estimation game was presented with novel climate change data, which is defined as climate change data and statistics that the students have not encountered before. This could typically be new climate change data since climate change data is evolving rapidly (Thacker, 2024). This was more effective in improving knowledge than traditional expository text. It is interesting to note that visualization of this data did not lead to measurable differences.

There is also competing evidence regarding how much instructional time is required to meaningfully build related behaviors. On one hand, short interventions may not be sufficient to build climate knowledge. For example, a one-hour teaching session on air pollution and climate change delivered to middle school students in the US did not yield measurable gains in knowledge (Abrham et al., 2023). Yet, despite this superficial exposure, students did report intentions to adopt more environmentally friendly behavioral choices such as taking public transportation.

This pattern contradicts the assumption that substantial knowledge must always precede any behavioral shifts (OECD, 2022). For example, evidence from a whole school-year program intervention paints a different picture. The CO_2 League program, implemented over a full school year across Czech Republic demonstrated that this in-depth intervention was successful in building climate change knowledge, which became a foundational role for building concern, self-efficacy, and willingness to act (Kolenatý et al., 2022). While the program does not establish causality, it suggests that sustained instruction strengthens the durability of knowledge, concern, and action since students have sufficient time to explore the complexities of climate systems.

While these two studies may seem contradictory, the CO_2 League program produced changes that extended far beyond personal, low-cost behaviors: students reported that the program exposed them to larger-scale climate mitigation and adaptation measures, motivated them to extend their engagement beyond individual lifestyle choices, and enabled them to participate in their school's municipality's climate planning processes. Alongside these systemic actions, the program also fostered a wide range of personal behavior changes. As noted by García-Vinuesa et al. (2024) and Lee et al. (2020), students often default to low-cost, low-disruption behaviors. In contrast, the one-hour lesson produced reported behavioral intentions in some domains that involved more personal sacrifice. This suggests that short exposures may be sufficient to shift attitudes temporarily, particularly around straightforward or personally resonant actions, but they may not provide the cognitive foundation needed to sustain or expand these actions over time, especially with systemic climate action.

Overall, building robust climate knowledge requires foundational scientific literacy. These foundations can be strengthened through the use of novel climate data and explicitly addressing misconceptions. Ideally, long-duration programs create the strongest gains in knowledge, concern, and personal and systemic forms of action. Yet most school systems face constraints that make this difficult to implement at scale. Recognizing these limitations, short-targeted interventions can still play a complementary role, especially when they introduce key concepts, spark curiosity, or motivate early pro-climate dispositions.

6.2.2 Increasing Learner Engagement

Another common factor in quality climate change education is leveraging strategies that increase learner engagement. Consistent themes emerge from empirical evidence: making learning personally relevant and using active engagement.

Monroe et al. (2019), in their systematic review, emphasized that making climate change personally relevant is effective given its spatial and temporal complexity. Their review highlights studies ranging from coastal India to Oregon in which students engaged with local issues, use geographic-specific climate data, and explore community-level vulnerabilities. More recent research extends this insight. In an undergraduate course in Taiwan, a futures-focused methodology was used to build personal relevance. Students first analyzed the current landscape of climate action in Taiwan, then examined plausible future scenarios based on scientific evidence, and finally developed narratives to communicate about those scenarios. This approach led to significant gains in knowledge and action competencies. Future-focused scenarios directly ties students to their own communities making learning personally meaningful and contextually relevant. While there are limitations to its generalizability as this study was based on only 34 undergraduate students, it shows a new, innovative approach in developing relevancy (Liu, 2024).

Another way to increase learner engagement is through interactive and multimodal learning experiences; in other words, students learn more when they actively interact with climate concepts rather than passively receiving information. Cumulative evidence about environmental education has long argued that active learning is a high-impact lever (Monroe et al., 2019). Immersive virtual learning is an example of effective active pedagogy to increase climate knowledge, self-efficacy, and behavioral intent (Stevenson et al., 2018). Because immersive virtual learning is not accessible to all schools, evidence on other forms of active pedagogy is particularly important. For example, a multi-modal climate change and food waste program in New Jersey was tested on 162 middle school students in a randomized control trial. This six-lesson program not only incorporated informative videos and slides to teach about relevant concepts, but every lesson had a hands-on activity like an e-learning game or card game where students were actively learning. This led to significant gains in knowledge, social norms, behavioral intentions, and perceived behavioral control, but not attitudes, self-efficacy, motivation to comply, or climate-friendly behaviors post-test (Elnakib et al., 2024). As seen with the CO_2 League program, these latter shifts may require longer durations. However, other techniques, that will be discussed in the subsequent section, may have strengthened this program, such as the integration of empowered action to bridge behavioral intent with actual behavior.

Lastly, Monroe et al. (2019) highlights that providing students with opportunities to apply climate action within their schools or communities is a powerful lever for climate change education. The studies they reviewed ranged widely in scale, from communicating and sharing information with their parents to an international exchange project. This resonates with research delineated about. For example, the CO_2 League program ends with an activity with students presenting their proposals with local and national stakeholders (Kolenatý et al., 2022). More recent research demonstrates that project-based learning enhances student knowledge, engagement, and environmental attitudes across different grade levels and contexts (Dominggus Peny et al., 2022; Kundariati et al., 2025; Maričić & Petković, 2022). Project-based learning is especially well suited to climate change because it naturally accommodates the interdisciplinary and real-world nature of the topic.

This lends climate change instruction naturally to an interdisciplinary approach as humanities classes such as civics, art, and English often provide additional and sometimes more flexible opportunities for projects and engagement within the structures of traditional school. In addition, interdisciplinary approaches to climate change education have also shown to be effective. At the university level, multiple studies have demonstrated that integrating climate content across disciplines improves student understanding, systems thinking, and engagement (Briguglio & Moncada, 2019; Hergert et al., 2010). Siegner and Stapert (2020) extend this evidence to the K–12 context, showing that interdisciplinary, place-based climate instruction in middle schools can increase climate literacy.

In summary, across studies, effective climate change education integrates core scientific literacy, engaging and personally relevant learning, and opportunities for students to apply climate action in real-world contexts. While short interventions can spark interest or influence behavioral intentions, longer and more immersive programs are most successful in producing durable gains in knowledge, concern, and meaningful action. Therefore, not only are interdisciplinary approaches effective, but may provide more opportunities for projects as well as longer climate change instruction.

Emerging developments deploying Artificial Intelligence in education suggests that AI could be used as an enabling tool that strengthens curriculum, pedagogy, assessment, and teacher capacity at scale. In practical terms, AI can help education systems address several of the constraints highlighted in this chapter. First, AI can function as a teacher-facing curriculum and lesson design partner that reduces preparation burdens and supports more engaging, place-based instruction. For example, teachers can use AI tools to draft interdisciplinary lesson plans aligned with core climate concepts (e.g., weather vs. climate; carbon cycle; adaptation and mitigation), generate locally relevant examples, create short formative assessments, and translate or adapt materials into students' home languages. This directly supports two of the most consistent principles in the climate education literature, personal relevance and active, engaging pedagogy, by making it easier for teachers to connect climate content to local livelihoods, hazards, and community priorities while expanding the variety of activities students can do.

Second, AI can support student learning through personalization and misconception diagnosis, which is especially important in classrooms where learning levels vary widely. AI-enabled tutoring (whether embedded in learning platforms or guided use of general-purpose language models) can provide scaffolded explanations, targeted practice, and immediate feedback; it can also help surface and correct misconceptions (e.g., confusing the ozone hole with climate change, or treating single cold days as evidence against warming). Used well, this can strengthen foundational climate literacy while also supporting metacognitive skills (planning, self-monitoring, and error detection) that help learners manage complexity and uncertainty. Tools such as Khan Academy's Khanmigo, MagicSchool AI, Brisk Teaching, and Diffit illustrate how AI can support differentiation and multilingual access.

Third, AI can strengthen climate education by expanding opportunities for authentic inquiry with real data, allowing students to learn climate change by "doing" rather than only reading. With appropriate guidance, students can use AI-supported tools to analyze local temperature and rainfall trends, compare emissions sources, interpret graphs and uncertainty, and build evidence-based explanations. This can also support deliberative discussion and civic reasoning, for example by helping students structure arguments, compare policy options, and evaluate claims, while requiring that students justify conclusions using credible sources and data.

Fourth, AI can make it easier to implement school- and community-based climate projects that translate learning into action (a consistent predictor of deeper engagement). AI tools can help students design and communicate projects such as campus energy or waste audits, heat-risk mapping, tree-canopy or shade surveys, or community interviews about climate impacts, producing summaries, visualizations, and presentations for local stakeholders. In these applications, AI can lower "transaction costs" (planning, drafting, analysis) so that more instructional time is spent on inquiry, collaboration, and decision-making.

However, these applications require careful attention to accuracy, bias, data privacy, and over-reliance on generative outputs, reinforcing the need for strong teacher professional development.

6.3 Frameworks for Climate Literacy

While the previous section discussed *how* climate change should be taught, it is also important to discern *what* students should learn. A growing set of climate and sustainability frameworks, produced by scientific bodies, international agencies, governments, higher education institutions, and non-governmental organizations, offer structured guidance about essential knowledge, skills, and attitudes. These frameworks vary in scope: some focus primarily on scientific literacy, others integrate competencies and values, and few focus on justice, local knowledge, and sociopolitical transformation. Taken together, they illustrate an emerging consensus: effective climate change education must combine understanding of climate systems with the ability to navigate social, political, and ethical dimensions of climate action.

Chapter 2 traced these dominant narratives about the purposes of climate change education, ranging form scientific literacy to critical and decolonial transformation. This section shifts from a narrative analysis of the purposes of climate change education to a more focused examination of the competencies students are expected to develop.

6.3.1 Science-Centered Climate Literacy Frameworks

The Intergovernmental Panel on Climate Change (IPCC) remains the most authoritative global body synthesizing climate science. Although not designed specifically for educators, the AR Working Group II Report provides an outline of climate impacts and adaptation challenges across sectors such as ecosystems, water, food systems, cities and infrastructure, health, and livelihoods as well as region-specific analyses (IPCC, 2022). For education, the IPCC's core value lies in its ability to provide validated, consensus-based scientific content that can anchor curriculum to the latest climate science and can connect climate change to real-world consequences and decision-making.

The *Climate Literacy: Essential Principles for Understanding and Addressing Climate Change* from the US Global Change Research Paper (USGCRP) synthesizes these important topics in a guide to teaching about climate change. This guide provides a climate literacy framework around eight essential principles:

1. Scientists understand the climate system through interdisciplinary observations and modeling.
2. Greenhouse gases shape Earth's climate.
3. Burning of fossil fuels and other human activities are causing the planet to warm.
4. Rapid warming and other large-scale climate changes threaten human and ecological systems.
5. Climate justice is possible if climate actions are equitable.
6. Humans can adapt social, built, and natural environments to better withstand the impacts of climate change.
7. Reducing emissions of greenhouse gases from human activities to net zero by 2050 can help limit global warming and climate change impacts.
8. A livable and sustainable future for all is possible with rapid, just, and transformational climate action.

In contrast to earlier versions, the 2024 update explicitly highlights Indigenous and local knowledge, emphasizes that a sustainable future depends on equitable action, and focuses on climate action to elicit hope. This marks a shift from purely knowledge-based, aligning more closely with global sustainability frameworks, which will be synthesized subsequently.

6.3.2 *Sustainability and Competency Frameworks*

Beyond scientific knowledge, climate change requires learning frameworks that cultivate systems thinking, ethical reasoning, collaboration, and the ability to act in uncertain contexts. Two influential frameworks, UNESCO's Education for Sustainable Development (ESD) and the European Commission's GreenComp, represent this shift towards competencies.

UNESCO's (2017) *Education for Sustainable Development Goals: Learning Objectives* positions education as central to achieving the 2030 Sustainable Development Goals (SDGs) and provides learning outcomes based on each SDG. Each SDG-centered learning objectives are categorized by cognitive, socio-emotional, and behavioral domains, emphasizing the need for problem-based and participatory learning. For example, when focusing on SDG 13: Climate Action, the learning objectives provide a framework for climate change education:

- Cognitive learning objectives:
  1. The learner understands the greenhouse effect as a natural phenomenon caused by an insulating layer of greenhouse gases.
  2. The learner understands the current climate change as an anthropogenic phenomenon resulting from increased greenhouse gas emissions.
  3. The learner knows which human activities—on a global, national, local and individual level—contribute most to climate change.
  4. The learner knows about the main ecological, social, cultural and economic consequences of climate change locally, nationally and globally and understands how these can themselves become catalysing, reinforcing factors for climate change.
  5. The learner knows about prevention, mitigation and adaptation strategies at different levels (global to individual) and for different contexts and their connections with disaster response and disaster risk reduction.
- Socio-emotional learning objectives:
  1. The learner is able to explain ecosystem dynamics and the environmental, social, economic and ethical impact of climate change.
  2. The learner is able to encourage others to protect the climate.
  3. The learner is able to collaborate with others and to develop commonly agreed-upon strategies to deal with climate change.
  4. The learner is able to understand their personal impact on the world's climate, from a local to a global perspective.
  5. The learner is able to recognize that the protection of the global climate is an essential task for everyone and that we need to completely re-evaluate our worldview and everyday behaviours in light of this.
- Behavioral learning objectives

1. The learner is able to evaluate whether their private and job activities are climate friendly and—where not—to revise them.
2. The learner is able to act in favour of people threatened by climate change.
3. The learner is able to anticipate, estimate and assess the impact of personal, local and national decisions or activities on other people and world regions.
4. The learner is able to promote climate-protecting public policies.
5. The learner is able to support climate-friendly economic activities.

In this particular framework, SDG 13: Climate Action is situated in the context of all the other SDGs, thereby integrating climate change with poverty, inequality, peace, consumption patterns, global citizenships, etc. This gives educators a roadmap for integrating climate within the broader sustainability goals. Further, it also includes key cross-cutting competencies such as systems thinking, anticipatory thinking, normative reasoning, strategic action, collaboration, and self-awareness, signaling a shift towards competencies rather than knowledge and facts alone.

The European Commission's GreenComp builds on and complements the UNESCO's SDG framework with their own articulated set of sustainability competencies based on their definition of sustainability: "prioritising the needs of all life forms and of the planet by ensuring that human activity does not exceed planetary boundaries" (Joint Research Centre of the European Commission, 2022). These learning competencies are separated into four domains:

1. Embodying sustainability values
    a. Valuing sustainability
    b. Supporting fairness
    c. Promoting nature
2. Embracing complexity in sustainability
    a. Systems thinking
    b. Critical thinking
    c. Problem framing
3. Envisioning sustainable futures
    a. Future literacy
    b. Adaptability
    c. Exploratory thinking
4. Acting for sustainability
    a. Political agency
    b. Collective action
    c. Individual initiative

Similar to UNESCO (2017), the sub-domains are separated into knowledge, skills, and attitude competencies. However, the knowledge domain is less focused on building scientific climate literacy.

6.3.3 *Justice and Context-Centered Frameworks*

As discussed in Chap. 2 the critical and decolonial approaches challenge the epistemic assumptions of mainstream climate education. Similarly, a third set of frameworks centers social justice, local knowledge, structural inequality, and the lived experiences of communities at the frontlines of climate impacts, often from the Global South. Dutta (2022) does not necessarily provide learning standards, but argues that current climate education frameworks are not grounded in justice-oriented approaches that connect climate change to local realities and grassroots movements. Therefore, they argue that climate change education needs to explicitly address:

- Historical responsibility and global inequality.
- Local concerns such as water scarcity, air pollution, land rights, and livelihoods.
- Structural injustices embedded in traditional climate solutions (i.e. renewable energy projects that displace Indigenous communities.
- Knowledge systems and struggles of marginalized peoples.
- Creative imagining to build just futures.

6.3.4 *Future-Ready Skills Frameworks*

Another body of work argues that climate education should be part of the broader transformation toward "future-ready" learning. As referenced in Chap. 2, Kwauk & Casey (2022) provide a framework on three categorizations of skills that are needed to transform unjust social and economic structures in a changing climate:

– Skills for Green Jobs: "Skills aimed at fulfilling the requirements of green jobs and supporting the transition to a low-carbon green economy," like data analysis, environmental and ecosystem management, marketing skills, etc.
– Green Life Skills: "Cross-cutting skills that serve both technical, instrumental, and adaptive, transformative ends," like adaptability, collaboration, critical thinking and reasoning, leadership, problem-solving, etc.
– Skills for a Green Transformation: "Adaptive skills aimed at transforming unjust social and economic structures," like coalition building, future and anticipatory thinking, political agency, respecting diverse viewpoints, working with complexity.

Another compelling model also illustrating this shift of integrating climate education into future life skills is from the "Future-Ready Classrooms" initiative in Uganda (HundrED, 2025). They are integrating climate literacy with digital literacy, coding, and AI, thereby situating climate problem-solving within a broader set of future-relevant skills. Further, this integrates other frameworks as these future life skills are directed towards community projects, preparing young people to participate in shaping their own futures within their local communities.

Sustainability learning objectives have also been integrated in the European Commission and the Organization for Economic Cooperation and Development (OECD)'s AILit Framework. For example, one of the listed competencies include "describe how AI systems consume energy and natural resources" where "learners explore the environmental impact of AI, including its energy and data infrastructure, and consider how responsible design and use can support sustainability" (OECD, 2025a). However, they do not explicitly incorporate other competencies related to sustainability; it merely shows how climate literacy can be embedded with other future life skills.

6.4 A Synthesized Framework for Climate Literacy

Clearly, a wide range of climate education frameworks exist. However, no matter how comprehensive a framework is, it must ultimately operate within the constraints of a school system, which includes competing resources, limited time, and continual pressure to integrate new content areas. As a result, a synthesized framework for climate education should be flexible and dynamic rather than fixed, shifting based on emerging climate realities and existing gaps within student knowledge, attitudes, and behavior towards climate.

As this chapter has shown, students lack the capacity to think systemically about climate, seldom understanding the social, economic, and health-related impacts of climate change. There is also a gap in understanding the most impactful solutions as well as how individual, community, and systemic actions interact. Lastly, students lack the scientific foundations and understanding behind the scientific concepts behind climate change, thereby leading to many misconceptions such as climate and weather, the relationship between the ozone layer and climate change, as well as acid rain, seismic activity, and pollution. While most students believe in climate change, there are varying levels of perceived self-efficacy and sense of responsibility. Therefore, there is an attitude-action gap: most students engage in low-impact individual actions such as recycling.

Across all the frameworks, several recurring themes emerge:

1. Scientific foundation—understanding climate systems, human drivers, and impacts are essential across all frameworks.
2. Systems thinking and interconnectedness—students need to connect social, economic, ecological, and political dimensions of climate.
3. Ethical and justice-oriented awareness—equity, historical responsibility, and local context are highlighted in justice-focused and updated science frameworks.
4. Action and agency—competencies for individual and collective action, from low-impact behaviors to systemic transformation, appear across most frameworks.
5. Future-ready skills—adaptability, collaboration, critical thinking, and digital & AI literacy are increasingly incorporated into climate education frameworks.

This framework is intentionally flexible and can be leveraged to bridge the gap between policy intent and classroom practice, as highlighted in Chap. 3. As Chap. 3 highlighted, case studies such as Mexico demonstrate that clearly identifying where climate content sits within curricula can catalyze resources and support, but a sole focus on content risks neglecting socio-emotional learning and action competencies, which are central to climate literacy.

Therefore, this framework could be used to

- Identify where systemic and interdisciplinary climate concepts fit across subjects and their learning standards.
- Integrate action-oriented competencies through community projects and cross-cutting learning experiences that already exist.
- Connect climate education with future-ready skills, redefining 21st-century competencies to include ethical reasoning, social responsibility, and transformative action.

In this way, the framework not only supports understanding but ensures students develop the knowledge, skills, and agency to act on climate challenges, laying the groundwork for curriculum design, assessment, and professional learning initiatives explored in the next chapter.

6.5 From Frameworks to Teaching

The OECD's recent Teaching Compass document usefully extends the Learning Compass vision from students to teachers and is particularly relevant for climate and environmental education. While the Learning Compass 2030 frames education around three interconnected dimensions of well-being, individual, collective, and environmental, the Teaching Compass argues that these aspirations "cannot be realised without corresponding changes in teaching practice and in the support systems that sustain teachers" (OECD, 2019, 2025b). In other words, if students are to develop the knowledge, skills, values, and agency needed to navigate climate change, teachers themselves must be equipped and recognised as agents of curriculum change in this domain.

Environmental well-being is explicitly named in the Teaching Compass as part of "the future we want," alongside individual and collective well-being. It is defined in terms of sustainability, environmental stewardship, and responsible resource use, with the warning that "without a stable and thriving planet…the foundations for both individual and collective well-being are fundamentally compromised" (OECD, 2025b, p. 8). From this perspective, climate change is not a peripheral topic but a core justification for rethinking educational purposes. The document links this to a call for "whole-system" thinking, urging educators to move beyond false binaries (for example, treating economic development and environmental protection as mutually exclusive) and to help learners understand how ecological, social, and economic systems are intertwined (OECD, 2025b, pp. 8–10). This is precisely the form of systems thinking that current climate literacy frameworks identify as essential.

The Teaching Compass also specifies what this implies for teacher knowledge and competencies. Under "knowledge about the world," it names climate change and sustainability, alongside migration, digitalization, and labor market shifts, as key global challenges that teachers must be able to connect to curriculum content in meaningful ways (OECD, 2025b, pp. 25–26). It emphasizes interdisciplinary and epistemic knowledge: teachers need to understand how climate knowledge is constructed and debated across disciplines, and to help students see not only what we know about climate change, but how that knowledge is generated, revised, and contested. This is complemented by a broad skills profile (critical thinking, problem-solving, scientific thinking, socio-emotional skills such as empathy and collaboration) and a values orientation that includes justice, global solidarity, and environmental awareness as "human or universal" values teachers should embody and model (OECD, 2025b, pp. 27–29). Together, these components describe the professional capacity required to teach climate change as both a scientific and an ethical issue.

Finally, the Teaching Compass situates climate and environmental education within teacher agency and co-agency. Teachers are described as "life integrators" who connect curriculum with students' everyday experiences and with real-world challenges like climate change, and as central agents in leading curriculum transformation toward sustainability (OECD, 2025b, pp. 17–18). Co-agency, with students, colleagues, families, communities, and increasingly AI systems, is presented as essential for tackling large-scale problems, including climate crises that directly affect school communities. At the same time, the framework underscores that such work is only sustainable if teacher well-being, physical, cognitive, and socio-emotional, is protected and strengthened (OECD, 2025b, pp. 30–32). For climate change education, this implies that what students can know and do will be tightly coupled to whether teachers have the knowledge, agency, support, and working conditions needed to engage climate issues deeply and consistently, rather than treating them as occasional add-ons. The next chapter examines the systemic conditions necessary to support teachers and students for effective climate change education.

References

Abrham, Y., Zeng, S., Tenney, R., Davidson, C., Yao, E., Kloth, C., Dalton, S., & Arjomandi, M. (2023). Effect of a single one-hour teaching session about environmental pollutants and climate change on the understanding and behavioral choices of adolescents: The BREATHE pilot randomized controlled trial. *PLoS ONE, 18*(11), Article e0291199. https://doi.org/10.1371/journal.pone.0291199

Aeschbach, V.M.-J., Schwichow, M., & Rieß, W. (2025). Effectiveness of climate change education—A meta-analysis. *Frontiers in Education, 10*, 1563816. https://doi.org/10.3389/feduc.2025.1563816

Bhattacharya, D., Carroll Steward, K., & Forbes, C. T. (2021). Empirical research on K-16 climate education: A systematic review of the literature. *Journal of Geoscience Education, 69*(3), 223–247. https://doi.org/10.1080/10899995.2020.1838848

Briguglio, L., & Moncada, S. (2019). The benefits and downsides of multidisciplinary education relating to climate change. In W. Leal Filho & S. L. Hemstock (Eds.), *Climate change and the role of education* (pp. 169–187). Springer International Publishing. https://doi.org/10.1007/978-3-030-32898-6_10

Dominggus Peny, A. A., Widodo, A., & Solihat, R. (2022). Project-based learning and problem-based learning in improving climate change literacy of junior high school students in disaster-prone areas. *Jurnal Pendidikan MIPA, 23*(2), 692–700. https://doi.org/10.23960/jpmipa/v23i2.pp692-700

Dutta, D. (2022). Environmental education for climate justice: An Indian perspective. In *Oxford research encyclopedia of education*. Oxford University Press. https://doi.org/10.1093/acrefore/9780190264093.013.1731

Elnakib, S., Subhit, S., Shukaitis, J., Rowe, A., Cava, J., & Quick, V. (2024). New Jersey leaves no bite behind: A climate change and food waste curriculum intervention for adolescents in the United States. *International Journal of Environmental Research and Public Health, 21*(4), 437. https://doi.org/10.3390/ijerph21040437

García-Vinuesa, A., Meira-Cartea, P. Á., & Caride-Gómez, J. A. (2024). Climate change education and secondary school students: A meta-synthesis (1993–2017). *Education in the Knowledge Society (EKS), 25*, e31358. https://doi.org/10.14201/eks.31358

Hergert, R., Barth, V., & Klenke, T. (2010). Interdisciplinary and interfaculty approaches in higher education capable of permeating the complexity of climate change. In W. Leal Filho (Ed.), *Universities and Climate Change* (pp. 107–115). Springer Berlin Heidelberg. https://doi.org/10.1007/978-3-642-10751-1_9

HundrED. (2025, September 29). *Future-ready classrooms: Digital & climate literacy*. https://hundred.org/en/innovations/future-ready-classrooms-digital-climate-literacy

IPCC. (2022). Pörtner, H.-O., Roberts, D. C., Tignor, M., Poloczanska, E. S., Mintenbeck, K., Alegría, A., Craig, M., Langsdorf, S., Löschke, S., Möller, V., Okem, A., & Rama, B. (Eds.), *Climate Change 2022: Impacts, adaptation, and vulnerability. Contribution of Working Group II to the Sixth Assessment Report of the Intergovernmental Panel on Climate Change*. https://doi.org/10.1017/9781009325844

Isac, M. M., Sandoval-Hernández, A., & Sass, W. (Eds.). (2025). *Knowledge and willingness to act pro-environmentally: Perspectives from IEA TIMSS 2019 and ICCS 2016 data*. Springer.

Joint Research Centre (European Commission). (2022). *GreenComp, the European sustainability competence framework*. Publications Office of the European Union. https://data.europa.eu/doi/10.2760/13286

Kolenatý, M., Kroufek, R., & Činčera, J. (2022). What triggers climate action: The impact of a climate change education program on students' climate literacy and their willingness to act. *Sustainability, 14*(16), 10365. https://doi.org/10.3390/su141610365

Kundariati, M., Ibrohim, I., Rohman, F., & Nida, S. (2025). Navigating climate change education: A problem-oriented project-based learning in secondary schools. *Science Activities, 62*(2), 156–170. https://doi.org/10.1080/00368121.2025.2463653

Kwauk, C. T., & Casey, O. M. (2022). A green skills framework for climate action, gender empowerment, and climate justice. *Development Policy Review, 40*(S2), Article e12624. https://doi.org/10.1111/dpr.12624

Lee, K., Gjersoe, N., O'Neill, S., & Barnett, J. (2020). Youth perceptions of climate change: A narrative synthesis. *Wires Climate Change, 11*(3), Article e641. https://doi.org/10.1002/wcc.641

Liu, S.-C. (2024). Exploring and narrating futures in undergraduate climate change education: An innovative teaching approach. *International Journal of Sustainability in Higher Education, 25*(9), 419–436. https://doi.org/10.1108/IJSHE-08-2023-0336

Maričić, O., & Petković, I. (2022). The effects of project-based learning about climate change. *Nastava i Vaspitanje, 71*(1), 109–124. https://doi.org/10.5937/nasvas2201109M

Monroe, M. C., Plate, R. R., Oxarart, A., Bowers, A., & Chaves, W. A. (2019). Identifying effective climate change education strategies: A systematic review of the research. *Environmental Education Research, 25*(6), 791–812. https://doi.org/10.1080/13504622.2017.136084
Mullis, I. V. S., Martin, M. O., Foy, P., & Hooper, M. (2017). *PIRLS 2016 international results in reading*. TIMSS & PIRLS International Study Center, Boston College.
Mullis, I. V. S., Martin, M. O., Foy, P., Kelly, D. L., & Fishbein, B. (2020). *TIMSS 2019 international results in mathematics and science*. TIMSS & PIRLS International Study Center, Boston College.
OECD. (2019*). OECD future of education and skills 2030: OECD learning compass 2030—A series of concept notes*. OECD Publishing.
OECD. (2022). *Are students ready to take on environmental challenges?* PISA, OECD Publishing. https://doi.org/10.1787/8abe655c-en
OECD. (2025a). *Empowering learners for the age of AI: An AI literacy framework for primary and secondary education (Review draft)*. OECD. https://ailiteracyframework.org
OECD. (2025b). *OECD teaching compass: Reimagining teachers as agents of curriculum change* (OECD Education Working Papers, No. 123). OECD Publishing.
Prati, G., Tzankova, I., Albanesi, C. & Cicognani, E. (2022). Longitudinal predictors of perceived climate change importance and worry among Italian youths: A machine learning approach. *Sustainability 14*.
Reimers, F. M. (2020). *Educating students to improve the world*. Springer.
Reimers, F. M., Chopra, V., Chung, C. K., Higdon, J., & O'Donnell, E. B. (2016). *Empowering global citizens: A world course*. CreateSpace Independent Publishing Platform.
Schulz, W., Ainley, J., Fraillon, J., Losito, B., Agrusti, G., Damiani, V., & Friedman, T. (2025). *Education for citizenship in times of global challenge: IEA International Civic and Citizenship Education Study 2022 international report*. Springer.
Schulz, W., Ainley, J., Fraillon, J., Losito, B., Agrusti, G., & Friedman, T. (2018). *Becoming citizens in a changing world: IEA international civic and citizenship education study 2016 international report*. Springer.
Siegner, A., & Stapert, N. (2020). Climate change education in the humanities classroom: A case study of the Lowell school curriculum pilot. *Environmental Education Research, 26*(4), 511–531. https://doi.org/10.1080/13504622.2019.1607258
Stevenson, K. T., Nils Peterson, M., & Bondell, H. D. (2018). Developing a model of climate change behavior among adolescents. *Climatic Change, 151*(3–4), 589–603. https://doi.org/10.1007/s10584-018-2313-0
Thacker, I. (2024). Supporting secondary students' climate change learning and motivation using novel data and data visualizations. *Contemporary Educational Psychology, 78*, Article 102285. https://doi.org/10.1016/j.cedpsych.2024.102285
This Is Planet Ed. (2025). *Youth climate literacy: A 2025 national snapshot*. The Aspen Institute. https://www.thisisplaneted.org/resources/youth-climate-literacy-a-2025-national-snapshot
USGCRP. (2024). *Climate literacy: Essential principles for understanding and addressing climate change*. U.S. Global Change Research Program. https://doi.org/10.7930/clg2024
UNESCO. (2017). *Education for sustainable development goals: Learning objectives*. UNESCO. https://doi.org/10.54675/CGBA9153
UNESCO. (2021). *Reimagining our futures together: A new social contract for education*. UNESCO. https://doi.org/10.54675/ASRB4722
Wildbichler, S., Haagen-Schützenhöfer, C., & Schubatzky, T. (2025). Students' ideas about the scientific underpinnings of climate change: A systematic review of the literature. *Studies in Science Education, 61*(1), 117–169. https://doi.org/10.1080/03057267.2024.2395206

Open Access This chapter is licensed under the terms of the Creative Commons Attribution-NonCommercial-NoDerivatives 4.0 International License (http://creativecommons.org/licenses/by-nc-nd/4.0/), which permits any noncommercial use, sharing, distribution and reproduction in any medium or format, as long as you give appropriate credit to the original author(s) and the source, provide a link to the Creative Commons license and indicate if you modified the licensed material. You do not have permission under this license to share adapted material derived from this chapter or parts of it.

The images or other third party material in this chapter are included in the chapter's Creative Commons license, unless indicated otherwise in a credit line to the material. If material is not included in the chapter's Creative Commons license and your intended use is not permitted by statutory regulation or exceeds the permitted use, you will need to obtain permission directly from the copyright holder.

Chapter 7
Thinking Systemically and Multidimensionally About Climate Change Education

Earlier chapters have shown that the relationship between education and climate change is systemic and bidirectional. Climate change acts on education systems through first-order effects that disrupt learning such as damaged schools, heat-induced learning loss, interrupted instruction, climate-driven migration and poverty. Education responds through second-order effects that shape knowledge, behaviors, institutions and technologies relevant to mitigation, adaptation and climate justice.

Yet global reviews make clear that, in most education systems, climate-change education remains fragmented and weakly institutionalized. UNESCO finds that although "environment" or "sustainability" appear in most policy and curriculum documents, climate change and biodiversity are far less frequently mentioned, and where they are, treatment is superficial and almost entirely cognitive (UNESCO, 2017, 2021). Teacher education, assessment and governance rarely prioritize climate-related competencies. Green-skills strategies are only weakly connected to education sector plans (International Labor Organization [ILO], 2019; Sabarwal et al., 2024). As a result, as discussed in the previous chapter, students fail to gain scientific depth and hold persistent misconceptions about climate change, which contributes to an attitude-action gap where pro-environmental beliefs do not consistently translate into high-impact or sustained behavioral change.

Research on curriculum, teacher professional development (PD) and school leadership in climate change education broadly corroborate this picture. We reviewed 40 empirical studies on climate-change curriculum integration and found that most programs produce substantial gains in student knowledge and more positive attitudes, but that social dimensions and action-oriented content are comparatively rare, and long-term behavioral change is weakly evidenced. Similarly, we reviewed 40 studies of climate-PD programs and found that teachers' content knowledge and self-efficacy improve markedly, often with large effect sizes, yet sustained changes in classroom practice and student learning are inconsistent, especially when PD is brief or unsupported by follow-up and institutional alignment. Finally, we reviewed 40 studies of

© The Author(s) 2026

F. Reimers and M. Wang-Aghania, *Educating for a Climate Changed Future*,
https://doi.org/10.1007/978-3-032-23256-4_7

leadership for climate and sustainability education and found that principals and headteachers can play critical roles in curriculum integration, PD and stakeholder engagement, but that success depends heavily on supportive policy environments, adequate resources and distributed leadership rather than isolated champions.

Taken together, these patterns suggest that, with respect to climate-change education, as with education in other domains, including foundational skills, many systems are drifting into what the World Development Report 2018 calls low-learning traps, stable equilibria in which most children attend school but learn very little, because incentives, information and power relations are misaligned with learning (World Bank, 2018). In the climate domain we can speak of low-climate-learning traps: climate appears in high-level rhetoric and occasional curriculum references, but is largely absent from what is taught, assessed and supported in classrooms. The system does not "see" climate-related learning, does not reward it, and therefore reinforces its neglect.

Four strands of work are particularly helpful for understanding how to escape such traps in climate-change education:

- Systems and complexity perspectives, which emphasize that education systems are complex adaptive systems, characterized by feedback loops, non-linearities and emergent behavior (Holland, 1998; Meadows, 2008; Mitchell, 2009; Reimers et al., 2026).
- Policy-coherence frameworks, which argue that standards, curriculum, assessment, teacher preparation, leadership and finance must be aligned rather than contradictory if reforms are to affect practice (Darling-Hammond, 2010; Elmore, 1996; Organization for Economic Co-operation and Development [OECD], 2010, 2013).
- The low-learning-trap analysis and escape strategies set out in the World Development Report: shining a light on learning, building coalitions for learning, and nurturing innovation and iteration (World Bank, 2018).
- Reimers' five-perspective framework (cultural, psychological, professional, institutional, political), which emphasizes that educational change is a multidimensional process and must be designed with all these dimensions in mind (Reimers, 2020).

This chapter weaves these four strands together and integrates the evidence we reviewed on curriculum, PD and leadership. It argues that climate-change education must be approached as a problem of escaping low-learning traps in a complex system. Doing so requires a systemic, multi-perspective approach, informed by policy coherence and futures thinking, and underpinned by deliberate planning for implementation. Such systemic transformation requires a multidimensional approach. We conclude the chapter pondering whether Artificial Intelligence can help support greater systemic coherence in climate change education.

7.1 Systems, Coherence and Low-Learning Traps in Climate-Change Education

Education systems are composed of interdependent components, curriculum, assessment, teacher education, school leadership, funding, governance, nested within broader political, economic and cultural environments. As complexity theorists have shown, such systems are complex and adaptive: relationships between cause and effect are often non-linear; small interventions can, under some conditions, trigger disproportionately large impacts through feedback loops; and new patterns of behavior emerge from local interactions without central control (Holland, 1998; Meadows, 2008; Mitchell, 2009).

Policy-coherence work on instructional systems emphasizes that policies and practices across standards, curriculum, assessment, teacher development, leadership and resources should be mutually reinforcing rather than contradictory (Darling-Hammond, 2010; Elmore, 1996; OECD, 2010, 2013). Coherence has:

- Vertical dimensions, where national or global goals are translated into district priorities and classroom practices.
- Horizontal dimensions, where education policies are aligned with other sectors (environment, labor, health, finance) and with other education policy domains (curriculum, infrastructure, or goals such as workforce development or civic education).

In coherent systems, teachers and school leaders experience relatively consistent signals about what matters. In incoherent systems, they are pulled in multiple directions: toward exam scores in a narrow set of domains, toward enrollment and access targets, toward short-term political initiatives. In this situation, climate-change education is easily crowded out.

The World Development Report (World Bank, 2018) sharpens this picture by showing how such incoherence and weak information can generate low-learning traps. When learning outcomes are not measured or used, when actors are rewarded for visible inputs rather than for learning, and when vested interests benefit from the status quo, systems can settle into stable patterns of low performance. Reform attempts that do not alter these underlying incentives and information flows often "evaporate" before changing practice.

For climate change education, this analysis suggests that:

- Climate is rarely included in the core accountability signals (national assessments, school reviews, funding formulas).
- Teachers receive mixed or weak messages about climate's importance relative to other demands.
- No strong coalitions are pushing for climate-related learning to be a central outcome.
- Innovation in climate education often remains isolated in pilots that are never scaled or institutionalized.

The research we reviewed reinforce this diagnosis. In curriculum, most climate initiatives are short-lived projects or one-off units without structural embedding. In PD, one-time workshops boost teacher awareness but do not reliably alter instruction. In leadership, inspiring principals can transform a school temporarily, but in the absence of policy and resource backing, such changes are fragile.

Escaping low-climate-learning traps therefore requires:

- Shining a light on climate learning: making climate-related knowledge, skills and dispositions visible through meaningful assessments and public reporting.
- Building coalitions for climate learning: mobilizing ministries, teachers, students, parents, civil society, employers and environmental actors around shared goals.
- Nurturing innovation and iteration: designing climate education reforms as adaptive processes with feedback loops, not one-off curriculum changes (World Bank, 2018).

Reimers' (2020) five-dimensional model offers a framework to transform the equilibrium that sustains low-learning traps through dynamics that are simultaneously:

- Cultural: how climate and education's purposes are understood in society.
- Psychological: how students and adults learn, and how they respond emotionally to climate information.
- Professional: how teachers and leaders are prepared, supported and recognized.
- Institutional: how structures, policies and routines support or hinder climate learning.
- Political: how interests and power shape what is prioritized and protected.

Efforts to build coherent climate-change education and to escape low-learning traps that address all five dimensions are more likely to succeed than those that are fragmentary.

With this framing, we now examine the role of five critical elements of the education system: curriculum; assessment and accountability; teacher professional development; school leadership and whole-school approaches; and governance. For each of them we ask:

- How does the current structure shape climate-related practice and risk low-learning traps?
- How do culture, psychology, professionalism, institutions and politics interact?
- How might futures and scenarios help stress-test and design interventions?
- What would credible, coherent, adaptive implementation planning look like?

The goal is not to suggest that alignment and planning guarantee transformation; in complex systems, change is always partly emergent. Rather, our argument is that without such alignment, without seeing across the five perspectives, without paying attention to policy coherence and the information–coalitions–innovation triad, climate-change education will remain an aspirational idea rather than a lived practice.

7.2 Curriculum: Climate Change Inside an Overloaded, Multidimensional System

Curriculum, a topic examined in Chaps. 3 and 6 in relation to content and pedagogy, is where societies formalize what they want students to know and be able to do. There are several levels of curriculum:

- Learning standards—broad expectations (e.g., understanding greenhouse effects, evaluating adaptation strategies).
- Curriculum frameworks—blueprints that organize content into strands and specify scope and sequence.
- Curriculum plans and materials—syllabi, textbooks, teacher guides, suggested projects.
- Units and learning goals—lesson objectives and tasks.

There is also the familiar triad of intended, implemented and attained curriculum. Much of what is currently labeled "climate-change education" sits in the intended layer, brief references in science or geography frameworks, without systematic translation into materials, teaching practices or measurable outcomes. This is characteristic of a low-learning trap: climate appears on paper but not in learning.

7.2.1 Institutional and Cultural Misalignment

UNESCO's reviews of national curricula and education sector plans show that terms like "environment" and "sustainable development" appear frequently, but "climate change" appears in fewer than half of curriculum frameworks, and "biodiversity" in only about one fifth (UNESCO, 2017). References are usually brief, confined to a few subjects, and not elaborated into learning progressions or performance expectations (UNESCO, 2021).

Our review of 40 empirical studies on climate-change curriculum integration identifies three main integration models:

- Embedded integration: climate content inserted into existing subjects (especially science and geography). This was the most common and least disruptive approach, but often resulted in fragmented and shallow coverage, scattered across topics and years.
- Cross-curricular or interdisciplinary approaches: climate used as a connecting theme across subjects (e.g., humanities, arts, social studies). These approaches were more likely to address social dimensions, ethics and action competence, but required greater coordination and teacher capacity.
- Standalone courses or modules: dedicated climate or climate-and-health courses, mainly in upper secondary and higher education. These allowed depth, especially for complex topics (e.g., climate–health biomechanisms in medical curricula),

but faced chronic problems of curriculum overcrowding and competition with examinable subjects.

Across these models, scientific foundations, causes, mechanisms, impacts, dominated; social dimensions (justice, policy, economics), emotional processing, and mitigation/adaptation strategies were far less common. This mirrors UNESCO's finding that when climate appears, it is usually treated as a neutral scientific topic rather than a socio-political and ethical one.

Two institutional features drive this pattern. The first is overcrowding. Existing frameworks already attempt to cover a wide range of priorities, foundational literacies, STEM, digital competence, civic education, "21st-century skills", health, entrepreneurship. Climate content is often layered on top without reprioritizing; structurally, it becomes a weak signal in a noisy system. The second feature is fragmentation. Different units develop frameworks, textbooks and guides on different timelines, with limited climate expertise and weak coordination. Education for Sustainable Development commitments may be declared as "mainstreamed", but without cross-unit planning to embed climate concepts coherently.

From a policy-coherence perspective, this means standards, curriculum, assessment and teacher training send inconsistent messages about climate. From a low-learning-trap perspective, climate has not been built into the core steering mechanisms of the system (World Bank, 2018). Teachers and leaders, responding rationally to the signals they receive, prioritize what is examined and resourced.

Research on curriculum also underscores the importance of design features: programs that enabled interactive learning (e.g., inquiry, modelling, project-based work), linked to local contexts, and supported by external partnerships (universities, NGOs, government) were more likely to show robust knowledge gains and positive attitudes (Siegner & Stapert, 2019). But even in these stronger designs, sustained behavior change and long-term retention were less consistently documented, especially when climate content was peripheral to assessment or schooling trajectories.

Culturally, curriculum reflects contested understandings of what education is for. In some contexts, schools are expected to transmit existing economic and social orders; climate content that raises questions of justice, consumption or intergenerational rights can be seen as disruptive or "political". If these cultural expectations are not acknowledged and addressed, they act as balancing feedback loops that neutralize climate reforms.

A coherent, trap-escaping strategy would:

- Clarify climate-related standards and progressions across subjects and grades, including social and action dimensions.
- Align these standards with textbook development, teacher standards and assessment blueprints.
- Budget and plan explicitly for iterative development, piloting, revision and scaling
- Engage communities and stakeholders in deliberations about climate's place in education's purposes, building cultural legitimacy.

7.2.2 *Psychological Foundations: What and How Students Can Learn About Climate*

Psychologically, curriculum must be grounded in what is known about how students think, feel and learn about complex issues. Climate literacy involves:

- factual knowledge.
- understanding dynamic causal mechanisms.
- interpreting data and uncertainty.
- applying concepts to new contexts.
- and sustaining engagement in the face of distressing information.

Research on deeper learning and 21st-century competencies suggests that climate curriculum should:

- Start from students' lived experiences of weather, hazards and landscapes, progressively introducing more abstract ideas like feedback and tipping points (Pellegrino & Hilton, 2012).
- Use multiple representations and link new ideas to existing schemas (Deans for Impact, 2015).
- Provide extended opportunities for practice and transfer, e.g., long-term projects on local vulnerability and adaptation options.
- Integrate cognitive, socio-emotional and interpersonal aims, including critical thinking, collaboration, empathy and self-efficacy.

Interactive and experiential pedagogies, place-based learning, modelling with tools like EzGCM, role-playing, field work, community engagement, are consistently associated with stronger student outcomes than lecture-based approaches. For example, programs using global climate models in secondary science classes and place-based climate science in U.S. schools showed significant gains in both climate content and epistemic knowledge, while generative learning approaches (students creating infographics or media) improved elementary students' understanding (Emidy, 2018; Forbes et al., 2020).

At the same time, these studies highlight that emotional responses are complex: well-designed curricula can simultaneously increase climate concern and self-efficacy. This underscores the need to design sequences that support emotional processing, acknowledging anxiety and loss while building agency and constructive hope.

Foundational literacy and numeracy are prerequisites; without them, students cannot access climate information or engage in quantitative reasoning (Sabarwal et al., 2024). In many low-learning systems, basic skills remain weak after years of schooling (World Bank, 2018). If climate content is layered on top of such fragile foundations, it risks remaining superficial or deepening inequities.

Implementation planning at this level implies:

- Specifying developmental sequences for key climate concepts and practices (including social and emotional competencies).

- Allocating realistic instructional time.
- Designing materials and teacher supports that address both cognitive and emotional demands (e.g., eco-anxiety).
- Monitoring and iteratively adjusting these sequences based on evidence of what students actually learn and retain over time.

7.2.3 Professional Alignment: What Curriculum Implies for Teacher Roles

Professionally, new climate curriculum implies changes in what teachers need to know and do. They require:

- Content knowledge (basic climate science, local impacts, green transitions).
- Pedagogical skills (project-based learning, facilitating deliberation on contested issues, interdisciplinary teaching).
- Socio-emotional skills (holding space for grief, anxiety and hope).

Reimers (2020) shows that systems that have broadened their curricular purposes, toward global citizenship, social and emotional learning, or 21st-century skills, have only succeeded when they have made commensurate investments in teacher learning and professional capital.

When climate change curriculum is supported with PD, time for collaborative planning, and ready-to-use resources, it is more likely to be enacted as intended; where teachers are left to "add climate" into overcrowded schedules without support, even well-designed frameworks will have little impact. In addition, this additional time with structured incentives could provide educators the motivation to engage professional learning communities, given how effective informal networks of educators can be for professional development as discussed in Chap. 4.

For climate-change education, curriculum design and teacher-development planning must be aligned. For each new climate-related expectation, planners should specify the knowledge and skills teachers will need, the forms of PD and support (courses, coaching, communities of practice) that will build these, realistic timelines for uptake and how progress in teacher practice and student learning will be monitored and fed back into PD design. If these links are not made, climate content remains aspirational and systems remain in low-learning equilibria.

7.2.4 Political Choices Embedded in Curriculum

Curriculum is politically saturated. Decisions about including fossil fuels, consumption patterns, Indigenous land rights, climate migration or historical responsibility are contested. Nationalist, populist or industry interests may resist some framings of climate justice; in some contexts, even acknowledging anthropogenic climate change can be politically sensitive.

Reforms that are politically misaligned, even if technically sound, are vulnerable to reversal, as happened with Poland's post-1999 structural reforms when a new government capitalized on negative narratives (Wiśniewski & Zahorska, 2020). Climate-curriculum reforms that do not build coalitions and manage political risk can meet similar fates.

A politically astute implementation strategy would map stakeholders' interests and likely reactions to different climate framings, identify champions (e.g., youth movements, scientific bodies, business leaders in green sectors), sequence content, starting with less contentious aspects while building deliberative spaces around more contested ones and generate and share credible evidence about the benefits and feasibility of climate-curriculum innovations, supporting narrative change over time.

7.3 Assessment and Accountability: Aligning Powerful Signals with Complex Aims

Assessment and accountability are among the strongest feedback mechanisms in education systems. What is measured and rewarded shapes what is taught (World Bank, 2018). If climate-related competencies are absent from examinations and accountability frameworks, they will remain marginal in practice, regardless of curricular rhetoric.

7.3.1 Institutional Tensions and Low-Learning Dynamics

Most national exams and large-scale assessments focus on factual recall and basic cognitive skills in a small set of subjects. Climate topics, when present, appear as isolated questions. Very few systems assess competences such as systems thinking about climate, interpreting emissions scenarios, evaluating adaptation options or reasoning about justice.

Similarly, school-accountability frameworks rarely include indicators of:

- students' climate understanding.
- school resilience to hazards.
- participation in climate-related projects.

Policy-coherence frameworks would describe this as misalignment between goals and instruments (OECD, 2010, 2013). The WDR would describe it as a core mechanism of low-learning traps: systems claim to value certain outcomes but do not measure them, so incentives and effort flow elsewhere (World Bank, 2018).

Implementation planning for climate-aligned assessment cannot be reduced to adding a few climate items to tests. It requires a phased strategy to pilot new kinds of items and tasks (e.g., scenario-based problem-solving, project portfolios), build item banks and scoring rubrics, train test developers and examiners, revise exam specifications and school review frameworks over time, and communicate with teachers and the public so that changes are understood and can be incorporated into instruction without panic. For example, a promising recent development is the 'Seal of Climate Literacy', which is a high school diploma endorsement for graduating secondary school students in the United States. This was piloted in Colorado and is slowly getting adopted in other states in the United States. Rather than simply assessing knowledge, students are required to meet minimum requirements for graduation, including taking a science course, complete a course that integrates climate literacy, and complete an experiential learning project (The Seal of Climate Literacy, n.d.). It is important to knowledge is not tested due to resource constraints; however, this demonstrates a new model that provides incentives for students to develop the skills to address climate change in their community."

7.3.2 Cultural and Psychological Risks of Narrow Assessment

Culturally, high-stakes assessments enshrine particular conceptions of worthwhile learning. If only narrow academic outcomes "count," aims such as stewardship, solidarity or intergenerational responsibility remain invisible. In exam-centric cultures, this can make it difficult for teachers to devote time to climate projects or reflective discussion.

Psychologically, assessment shapes learners' identities and sense of agency. Narrow, high-pressure tests can signal that climate is peripheral, reduce climate to decontextualized facts or exacerbate anxiety without building capacity to act.

A trap-escaping strategy would introduce climate-related assessment in ways that broaden what is valued, support formative feedback and growth, not only sorting and are sequenced gradually so that schools can adapt.

7.3.3 Professional and Political Dimensions

Professionally, teachers' choices are powerfully shaped by what is assessed. Many have limited experience with designing and using performance tasks or projects that capture complex competences. Without targeted assessment-literacy support, climate-related tasks may be seen as risky or unmanageable.

Politically, assessment reform is sensitive. Changing exams to include climate tasks can trigger debates about ideology, fairness or standards. Accountability systems can be used to reward or punish schools, making educators wary of new indicators.

Implementation planning thus needs to include PD on rich assessment of climate competences, engage universities and employers in recognizing such competences and construct coalitions to support more holistic assessment, while maintaining public confidence.

7.4 Teacher Professional Development: Building Climate-Ready Professional Capital

Teachers are the proximate determinants of the instruction that students experience. UNESCO's surveys find that large shares of teachers feel ill-prepared to teach climate and environmental topics (Plutzer et al., 2016; UNESCO, 2021).

7.4.1 A Professional Perspective on Climate Education

Reimers' professional perspective emphasizes that instruction should be grounded in expert knowledge and that roles and institutions should enable professionals to use and generate such knowledge (Reimers, 2020). For climate-change education, this implies:

- strong foundations in relevant content.
- pedagogical repertoires for interdisciplinary, project-based and participatory climate learning.
- skills for facilitating emotionally charged discussions.
- and structures that support continuous learning from practice.

Comparative work on ambitious reforms, whether in global citizenship, social and emotional learning or 21st-century skills, shows that one-off workshops are ineffective; sustained, school-embedded PD linked to curriculum and assessment is needed (Reimers & Chung, 2016, 2018; Reimers, 2018).

A review of climate-change PD programs provides quantitative support for this argument. Across studies, teachers' climate-science content knowledge often increased dramatically, for example, from 38.1 to 75.6% on content assessments (normalized gains around 64%), or with average grade increases of 28% in large multi-state programs (Brey et al., 2015; Teed & Franco, 2014). Effect sizes for conceptual understanding were often large (e.g., $d = 1.32$ for physics pre-service teachers), and self-efficacy gains were substantial ($d \approx 0.75$).

Yet translation of these gains into sustained practice was uneven: many teachers developed new instructional units and integrated inquiry-based approaches; in some programs, over 90% reported applying their learning in the classroom (Reyes et al.,

2024). In other contexts, especially after brief PD, teachers reverted to traditional lecture and textbook-based methods, and climate topics remained rare or superficial in their teaching (Mavuso et al., 2022).

Where student outcomes were measured, gains, though positive, were more modest than teacher-level gains (e.g., average normalized student learning gains around 56%, moderate effect sizes around d = 0.49), and students' reasoning about mitigation options and tangible actions often remained underdeveloped (Drewes et al., 2018).

In short, PD programs can reliably enhance teachers' knowledge and confidence, but these improvements do not automatically translate into proportional improvements in classroom practice and student learning.

Professional development focused on climate change education cannot be a technical fix without altering the institutional conditions that shape practice, time, curriculum, assessment, leadership and support. Escaping traps in the climate domain requires treating climate-related PD as a core function of the system and embedding it in a broader alignment effort.

7.4.2 *Psychological and Institutional Conditions for Teacher Learning*

Teachers bring prior beliefs and "immunities to change" (Kegan & Lahey, 2009). Some may see climate as "too political", feel overwhelmed by its complexity, or be anxious about discussing distressing futures with students. Concerns about climate as a controversial topic, and perceived complexity of integrating it into non-science subjects, are common barriers (Li et al., 2019; Sawaddee & Chiravate, 2025). Programs should explicitly address these anxieties, by modelling how to handle controversy, or by showing how climate fits naturally into English, social studies or arts.

Effective implementation plans for climate-related PD should therefore:

- diagnose teachers' starting points (knowledge, beliefs, concerns).
- design learning experiences that address both cognition and emotion, e.g., scenario workshops that help teachers explore possible futures and their educational implications.
- provide follow-up support through coaching, peer observation and communities of practice.
- explicitly connect PD to curriculum and assessment changes, so that teachers see climate as integral, not optional.

Design features of PD programs are associated with stronger and more sustained impacts. Extended duration with follow-up support, such as multi-week or multi-year programs with summer institutes plus school-year coaching, produces more durable changes than short workshops (Dupigny-Giroux et al., 2012; Ellins et al., 2014).

Hands-on, inquiry-based activities that engage teachers in authentic investigations (field work, lab experiments, remote sensing, data analysis) are more likely to shift their pedagogy toward inquiry (Ceyhan et al., 2025; Teed & Franco, 2014). Communities of practice and collaborative design in programs that fostered teacher networks, co-design of lessons, and reflective dialogue supported ongoing experimentation, while isolated training events did not (Allen & Crowley, 2014; Feldman et al., 2021). Finally, alignment with local curriculum and context is important. PD that helped teachers adapt climate topics to their existing syllabi and community realities (e.g., climate-smart agriculture in the Philippines, or Indigenous worldviews in Native American communities) was more readily enacted (Kerns et al., 2017; Manalo et al., 2016).

Institutionally, PD must be resourced (time, money, expertise) and embedded in teachers' work, not bolted on. Indicators such as the proportion of schools with ongoing climate-related PD, or the proportion of teachers implementing climate projects, can provide information to guide adaptation, echoing the WDR's emphasis on using data to support system learning (World Bank, 2018).

Politically, engaging teacher unions and professional associations or universities in the co-design of climate-PD initiatives can build ownership and reduce the risk of backlash.

7.5 School Leadership and Whole-School Approaches: Local System Integrators

Schools are systems in their own right, with cultures, routines, resource flows and external relationships. Principals and leadership teams sit at the intersection of policy and practice, making them central to aligning climate-change education at the local level. From an institutional lens, leaders translate national frameworks into school-level plans; allocate time and resources for climate learning and teacher collaboration; coordinate integration of climate themes across subjects; and oversee physical changes (e.g., greening school grounds, reducing carbon footprint) that can make schools sites of lived climate action.

From a professional lens, leaders can cultivate their schools as learning organizations (Kools & Stoll, 2017; Reimers, 2018). They model openness to learning and experimentation, support teacher leadership in climate education and build connections to external partners (NGOs, universities, local government).

Culturally, leaders shape narratives about why climate matters for their community and how it fits with local values. Politically, they navigate tensions and build local coalitions.

Distributed leadership, where responsibility for sustainability is shared among leadership teams, teachers, students and sometimes parents, is consistently associated with more robust and sustainable change than reliance on a single "green champion" (Leo & Wickenberg, 2013; Zala-Mezö et al., 2019).

Quantitative evidence, though limited, is suggestive. In Kenya, principals' incorporation of CCE into curriculum was statistically significantly associated with higher student climate-awareness levels (Debe & Chepkonga, 2025). In Swiss "eco-schools", more efficient leadership strategies correlated with perceived quality of school-wide ESD implementation (Zala-Mezö et al., 2019).

The policy environment shapes how schools can address climate change education. In countries where ESD or climate education was embedded in national frameworks and inspection regimes (e.g., Germany, parts of Scandinavia), principals reported clearer mandates and more stable support (Müller et al., 2021, 2022). In contexts with policy silence or inconsistency (e.g., England between 2010–2022, New Zealand's diffuse frameworks), leadership for climate was discretionary and fragile, heavily dependent on individual conviction and local networks (Bolstad, 2020; Higham et al., 2025; School leadership and climate change education, 2025).

Implementation planning at the school-leadership level should therefore:

- embed climate goals in school improvement plans, with clear objectives, roles, timelines and indicators.
- create structures for student participation (e.g., climate councils, project teams).
- map local stakeholders and design engagement strategies that are culturally and politically attuned.
- intentionally distribute leadership for climate across staff and students to avoid over-reliance on individuals.
- leverage regional and international networks (e.g., Eco-Schools, UNESCO ASPnet) as sources of expertise and legitimacy (Cebrián et al., 2022; Hargis et al., 2021).

If school leadership is neglected, climate initiatives risk remaining ad hoc, dependent on a few enthusiastic individuals, and vulnerable to staff turnover and external shocks, another pathway into a low-learning trap.

7.6 Governance: Multi-level Alignment and Policy Coherence in a Climate-Constrained Future

Finally, coherence in climate-change education depends on governance: the formal and informal arrangements through which decisions are made, resources allocated, and accountability exercised, both within education and between education and other sectors.

7.6.1 Institutional Architecture and Vertical Alignment

Globally, education operates within a normative environment shaped by UNESCO, UNICEF, the World Bank, regional organizations and international agreements such as the Paris Agreement and the 2030 Agenda for Sustainable Development (Meyer et al., 1992; UNESCO, 2021; United Nations, 2015). Within countries, ministries of education interact with ministries of environment, energy, labor, finance and planning, as well as sub-national governments, examination boards and civil society.

Vertical misalignment is common: global climate-education commitments are only thinly reflected in national sector plans, and even more weakly in sub-national implementation. Horizontal fragmentation is also common: environmental and skills policies are often poorly coordinated (ILO, 2019; Sabarwal et al., 2024).

From a policy-coherence and low-learning-trap perspective, this means that climate-learning goals lack clear institutional homes, explicit budgets and delivery chains and integrated monitoring. Principals in supportive policy environments can leverage national ESD mandates to legitimize climate work, while those in policy vacuums must improvise, negotiate competing demands and hope that successor leaders will not dismantle their efforts.

Implementation planning at this level must:

- embed concrete education components in national climate strategies (e.g., NDCs, adaptation plans).
- embed climate components in education sector plans.
- clarify responsibilities and funding flows across levels of government.
- establish regular reporting on climate-education initiatives and outcomes.

7.6.2 Political Interests, Culture and Bureaucratic Capacity

The WDR underlines that reforms are shaped by political interests and organizational capacities (World Bank, 2018). In climate-change education, similar dynamics apply. Some actors may resist climate-aligned reforms because they threaten economic interests or ideological positions. Ministries may lack staff with climate or systems expertise. Bureaucracies accustomed to compliance-oriented routines may struggle with adaptive implementation.

Reimers (2020) argues that systems must themselves become learning organizations, capable of diagnosing, experimenting and adapting. Governing climate-change education in a way that avoids low-learning traps requires investments in bureaucrats' professional knowledge of climate and complexity, institutionalization of cross-sector planning processes and political strategies to build and sustain coalitions for climate learning, even across electoral cycles. Scenario processes can help by stress-testing sector plans against plausible futures (e.g., more frequent disasters, major climate-related migration) and identifying strategies that are robust across them.

7.7 Can Artificial Intelligence Support Systemic Coherence in Climate Change Education?

Reimers et al.'s Systems Perspective on AI and education in the Global South is useful for thinking about coherence in climate change education because it starts from a familiar reform failure: ambitious curricular goals are routinely undermined by misaligned assessments, weak teacher support, and administrative routines that pull schools back toward rote instruction (Reimers et al., 2026). In climate change education, this misalignment is especially consequential: when climate-related competencies (systems thinking, data reasoning, civic deliberation, and action competence) are not embedded in the "core steering mechanisms" of the system: assessment blueprints, teacher development routines, school improvement plans, and governance dashboards, climate learning remains peripheral, episodic, and vulnerable to being crowded out by high-stakes priorities. AI cannot solve this problem on its own, but it can help re-engineer the information flows and feedback loops that make coherence possible.

First, AI can help systems "shine a light" on climate learning by making it visible and measurable in ways that are feasible at scale. A central bottleneck in many systems is that assessment remains dominated by high-stakes, recall-oriented examinations, even when curricula espouse deeper competencies. AI offers a pathway to shift assessment toward more formative, learning-oriented functions, continuous feedback, diagnosis of misconceptions, and performance-based tasks, rather than relying exclusively on episodic exams (Reimers et al., 2026). Applied to climate change education, this means AI-enabled tools could support (a) low-stakes checks of conceptual understanding (e.g., distinguishing weather from climate; interpreting uncertainty), (b) scenario-based tasks (e.g., evaluating adaptation options under constraints), and (c) portfolio-based evidence of sustained work (projects, investigations, civic products). The key coherence point is that if climate competencies become assessable in practical, lower-cost ways, they can be integrated into classroom routines and into accountability signals without forcing teachers into an impossible workload.

Second, AI can strengthen alignment between the intended, implemented, and achieved curriculum by tightening feedback loops from classrooms back to curriculum design. Education systems often lack mechanisms to learn, at scale, what students are actually mastering and where misconceptions persist. Aggregated learning data could support "curriculum intelligence hubs" that identify patterns and inform iterative curriculum improvement (Reimers et al., 2026). For climate change education, similar analytics could help ministries and curriculum agencies detect, for example, that students systematically misunderstand feedback loops, confuse mitigation with adaptation, or default to low-impact behavioral framings, then revise learning progressions, materials, and teacher supports accordingly. This is not merely a technical upgrade; it is a coherence mechanism that connects what is said in curriculum frameworks to what is learned in practice.

Third, AI can improve professional coherence by making climate-related teacher learning more continuous, job-embedded, and instructionally actionable. Teachers are the primary agents translating curriculum into learning, yet professional development is frequently brief, generic, and disconnected from classroom realities. AI tools can support teacher development in climate education in at least three coherence-supporting ways:

- Lesson planning and material generation aligned to standards: teacher-facing AI can generate localized, differentiated climate lessons and formative checks aligned to official curriculum, reducing the time–cost of implementation while increasing fidelity to intended goals (Reimers et al., 2026).
- Instructional coaching at scale: AI-enabled coaching (for example via low-bandwidth channels such as messaging platforms) can provide on-demand support for common instructional problems, addressing misconceptions, facilitating deliberation, integrating data activities, helping shift the implemented curriculum toward the intended one (Reimers et al., 2026).
- Assessment literacy for complex competencies: because climate education aims extend beyond recall, AI-supported tools can scaffold teachers in using rubrics, projects, and formative feedback, linking teacher learning directly to assessment reform (Reimers et al., 2026). The coherence implication is that teacher development becomes part of the same reform "bundle" as curriculum and assessment, rather than an optional add-on.

Fourth, AI can support institutional coherence through school leadership and management tools that re-center instructional leadership and collaboration. School leaders in many systems are overburdened by administrative work, leaving little time to build cultures of professional collaboration or to support instructional improvement. AI can help automate routine administrative tasks and provide dashboards that summarize climate-relevant indicators, such as participation in climate projects, integration across subjects, or patterns of student understanding, so leaders can treat climate education as an explicit school improvement priority rather than an invisible aspiration (Reimers et al., 2026). In addition, AI-supported professional learning communities can reduce the transaction costs of collaboration across schools (e.g., summarizing discussions, curating resources, and enabling co-design), which is especially important for interdisciplinary climate learning that no single teacher "owns" (Reimers et al., 2026).

Fifth, AI can enhance governance coherence by improving data use, communication, and resource allocation across levels of the system. Education governance in low-resource contexts is often constrained by weak data systems, limited analytic capacity, and slow feedback loops. AI, particularly through natural language interfaces and automated analytics, can make it easier for system actors to ask questions of their data ("Where are climate misconceptions concentrated by grade and region?" "Which schools are implementing climate projects, and with what supports?") and to shorten the cycle from information to action (Reimers et al., 2026). This matters for climate education because coherence frequently fails in the "middle layer" between national intention and school implementation: AI-enabled monitoring and support systems can help coordinate curriculum materials, PD provision, and assessment changes so that schools receive consistent signals and usable supports rather than fragmented initiatives.

7.8 Conclusion: Seeing Climate-Change Education Multidimensionally, in a Complex System, and Planning to Escape Low Learning Traps

Thinking systemically about climate-change education requires more than mapping components; it requires seeing how they interact through the lenses of culture, psychology, professionalism, institutions and politics (Reimers, 2020), and recognizing how those interactions can lock systems into low-climate-learning traps (World Bank, 2018).

Climate is integrated, embedded, cross-curricular or standalone, and how it is taught, interactive and contextualized versus didactic, shapes student knowledge and attitudes, without alignment to assessment, PD and leadership, behavior change and long-term impact will remain limited. Teacher knowledge and confidence can be dramatically improved, and sustained changes in practice require extended, inquiry-based, context-aligned PD and supportive school structures. Principals and headteachers can make climate and sustainability central to school purpose, but only when backed by policy, resources, cultural resonance and distributed leadership.

Policy-coherence frameworks remind us that curriculum, assessment, teacher development, leadership and governance must be mutually reinforcing rather than contradictory. Complexity science reminds us that such coherence is dynamic: it is achieved, if at all, through iterative cycles of experimentation, feedback, negotiation and revision (Holland, 1998; Meadows, 2008; Mitchell, 2009).

The low-learning-trap lens adds three imperatives:

- generate credible information about climate learning.
- build coalitions that care about it.
- embed innovation and iteration in reforms (World Bank, 2018).

Futures and scenarios help us ask: "Education for which climate futures, for whom, and how?" They can anchor curricular design, professional learning, school planning and policy dialogue in an honest confrontation with uncertainty, while supporting constructive agency.

But systemic and futures thinking will not change outcomes unless they are connected to deliberate planning for implementation. This means moving from high-level aspirations ("greening education") to specific legacy goals; diagnosing root causes of current gaps; selecting priority levers; designing coherent strategies across curriculum, assessment, teacher development, leadership and governance; and then specifying, in realistic detail, the sequence of activities, responsibilities, resources, timelines, indicators and assumptions that will carry those strategies into classrooms and communities. It also means treating these plans as living documents, revised as systems learn what works.

Coherence in climate-change education is not achieved by perfect top-down design, but by iteratively aligning curriculum, assessment, teacher development, leadership and governance so that they send consistent signals, create reinforcing feedback loops, avoid low-learning traps, and remain robust across a range of plausible climate and social futures. Without this multidimensional, systemic, future-oriented and implementation-conscious approach, well-intentioned climate initiatives will continue to be diluted by stronger, misaligned dynamics.

With such an approach, grounded in complexity science, policy coherence, Reimers' five perspectives, and the emerging evidence base on curriculum, PD and leadership, education systems can move closer to fulfilling their potential as engines of climate adaptation, mitigation and transformation, helping students not only understand a changing climate, but also become capable, ethical and hopeful agents in shaping the world they will inherit.

References

Allen, L. B., & Crowley, K. (2014). Challenging beliefs, practices, and content: How museum educators change. *Journal of Museum Education, 39*(2), 157–171.

Bolstad, R. (2020). *Climate change and sustainability in secondary schools: Findings from a 2020 survey of English-medium secondary and composite schools.* New Zealand Council for Educational Research.

Brey, J., Mills, E. W., Day-Miller, E., Geer, I., Weinbeck, R. S., Blair, B., Hopkins, E. J., O'Neill, K., Nugnes, K. A., & Asokan, A. (2015). The AMS DataStreme Earth's Climate System teacher professional development course: Results from a three-year study. *Bulletin of the American Meteorological Society, 96*(9), 1559–1570.

Cebrián, G., Mogas, C., Palau, R., & Fuentes, M. (2022). Sustainability and the 2030 Agenda within schools: A study of school principals' engagement and perceptions. *Environmental Education Research, 28*(2), 233–252. https://doi.org/10.1080/13504622.2021.1937058

Ceyhan, G., Budak, Ü. S., Gungor Cabbar, B., Ertugrul, N., & Genc, B. (2025). Using a systems thinking approach for a climate change education training program in teacher education. *Science Activities, 62*(1), 1–17.

Darling-Hammond, L. (2010). *The flat world and education: How America's commitment to equity will determine our future.* Teachers College Press.

Deans for Impact. (2015). *The science of learning.* Deans for Impact.

Debe, J. M., & Chepkonga, S. (2025). Influence of principals' incorporation of climate change education into curriculum on climate change awareness amongst students in public secondary schools in Westland Subcounty, Kenya. *International Journal of Current Science Research and Review, 8*(7), 3703–3708. https://doi.org/10.47191/ijcsrr/V8-i7-57

Drewes, A., Henderson, J. A., & Mouza, C. (2018). Professional development design considerations in climate change education: Teacher enactment and student learning. *Journal of Science Teacher Education, 29*(3), 210–229. https://doi.org/10.1080/1046560X.2018.1432228

Dupigny-Giroux, L.-A., Toolin, R., Hogan, S., & Fortney, M. (2012). The satellites, weather and climate (SWAC) teacher professional development program: Making the case for climate and geospatial literacy. *Journal of Geoscience Education, 60*(2), 103–117.

Ellins, K., Ledley, T., Haddad, N., McNeal, K., Gold, A., Lynds, S., & Libarkin, J. (2014). EarthLabs: Supporting teacher professional development to facilitate effective teaching of climate science. *Journal of Geoscience Education, 62*(3), 407–418.

Elmore, R. F. (1996). Getting to scale with good educational practice. *Harvard Educational Review, 66*(1), 1–26.

Emidy, M. (2018). *Changing climate in the classroom: A model for place-based climate science education in San Diego County* [Master's thesis]. University of San Diego.

Feldman, A., Nation, M., & Laux, K. (2021). The effects of extended action research-based professional development on the teaching of climate science. *Educational Action Research, 29*(5), 770–789.

Forbes, C. T., Chandler, M. A., Bhattacharya, D., Steward, K. C., Blake, J., Johnson, V., Morrow, M., Mason, W., & DeGrand, A. (2020). Fostering climate literacy with global climate models in secondary science classrooms. *Journal of Geoscience Education, 68*(2), 160–178.

Hargis, K., McKenzie, M., & LeVert-Chiasson, I. (2021). A whole institution approach to climate change education. In K. E. Winters, N. Walshe, & H. France (Eds.), *Curriculum and learning for climate action* (pp. 163–180). Routledge.

Higham, R., Kitson, A., & Sharp, S. M. (2025). 'What do I do? Save the environment or let children go hungry?' Leading English schools at time of climate crisis. *International Journal of Leadership in Education, 28*(3), 345–365.

Holland, J. H. (1998). *Emergence: From chaos to order*. Oxford University Press.

International Labour Organization. (2019). *Skills for a greener future: A global view based on 32 country studies*. ILO.

Kegan, R., & Lahey, L. L. (2009). *Immunity to change: How to overcome it and unlock potential in yourself and your organization.* Harvard Business Press.

Kerns, A., Honwad, S., & McLain, E. (2017). A culturally relevant teacher professional development for teaching climate change to Native American students. In K. Winslow & J. Thompson (Eds.), *Climate change education: A primer for K–12 educators* (pp. 215–229). Springer.

Kools, M., & Stoll, L. (2017). *What makes a school a learning organisation? A guide for policy makers, school leaders and teachers* (OECD Education Working Papers, No. 137). OECD Publishing.

Li, C., Monroe, M. C., Oxarart, A., & Ritchie, T. (2019). Building teachers' self-efficacy in teaching about climate change through educative curriculum and professional development. *Applied Environmental Education & Communication, 18*(3), 251–269. https://doi.org/10.1080/1533015X.2018.1454351

Leo, U., & Wickenberg, P. (2013). Professional norms in school leadership: Change efforts in implementation of education for sustainable development. *Journal of Educational Change, 14*(4), 403–422.

Manalo, J., Balmeo, K. P., Berto, J. C., Saludez, F. M., Villaflor, J., & Pagdanganan, A. M. (2016). Integrating climate-smart rice agriculture into secondary-level curriculum: Lessons from three high schools in the Philippines. *SpringerPlus, 5*, Article 984.

Mavuso, M., Olawumi, K., Khalo, X., Kafu-Quvane, B., & Mzilikazi, B. (2022). Implementation of teacher capacitation programs to integrate climate change education: The case study of Geography teaching in South African secondary schools. *International Journal of Learning, Teaching and Educational Research, 21*(12), 1–20.

Meadows, D. H. (2008). *Thinking in systems: A primer*. Chelsea Green.

Meyer, J. W., Ramirez, F. O., & Soysal, Y. N. (1992). World expansion of mass education, 1870–1980. *Sociology of Education, 65*(2), 128–149.

Mitchell, M. (2009). *Complexity: A guided tour*. Oxford University Press.

Müller, U., Hancock, D., Stricker, T., & Wang, C. (2021). Implementing ESD in schools: Perspectives of principals in Germany, Macau, and the USA. *Sustainability, 13*, Article 10180.

Müller, U., Hancock, D., Wang, C., Stricker, T., Cui, T., & Lambert, M. (2022). School leadership, education for sustainable development (ESD), and the impact of the COVID-19 pandemic: Perspectives of principals in China, Germany, and the USA. *Education Sciences, 12*(3), 203.

Organisation for Economic Co-operation and Development. (2010). *OECD reviews of human resource management in government: Brazil 2010*. OECD Publishing.
Organisation for Economic Co-operation and Development. (2013). *Synergies for better learning: An international perspective on evaluation and assessment*. OECD Publishing.
Pellegrino, J. W., & Hilton, M. L. (Eds.). (2012). *Education for life and work: Developing transferable knowledge and skills in the 21st century*. National Academies Press.
Plutzer, E., McCaffrey, M., Hannah, A. L., Rosenau, J., Berbeco, M., & Reid, A. H. (2016). Climate confusion among U.S. teachers. *Science, 351*(6274), 664–665.
Reimers, F. M. (2018). *Teaching and learning for the twenty-first century: Educational goals, policies, and curricula from six nations*. Harvard Education Press.
Reimers, F. M. (2020). *Educating students to improve the world*. Springer.
Reimers, F. M., Azim, Z., Palomo, M.-R., & Thony, C. (2026). *Artificial intelligence and education in the Global South: A systems perspective*. Springer. *Springer*. https://doi.org/10.1007/978-3-032-11449-5
Reimers, F. M., & Chung, C. K. (Eds.). (2016). *Teaching and learning for the twenty-first century: Educational goals, policies, and curricula from six nations*. Harvard Education Press.
Reimers, F. M., & Chung, C. K. (2018). *Preparing teachers to educate whole students: An international comparative study*. Harvard Education Press.
Reyes, P. C., Viviani, M., & Robles, C. M. (2024). Strengthening capacities in Latin America by designing teacher professional development programmes for climate change education. *Journal of Professional Capital and Community, 10*(1), 6–19. https://doi.org/10.1108/JPCC-12-2023-0098
Sabarwal, S., Venegas Marin, S., Spivack, M., & Ambasz, D. (2024). *Choosing our future: Education for climate action*. World Bank.
Sawaddee, N., & Chiravate, B. (2025). Investigating Thai local secondary school English teachers' perceptions of climate change and its classroom teaching. *Journal of Arts and Thai Studies, 47*(2). https://doi.org/10.69598/artssu.2025.4447
School leadership and climate change education. (2025). *School leadership and climate change education: A synthesis of empirical research* [Unpublished research report].
Siegner, A., & Stapert, N. (2019). Climate change education in the humanities classroom: A case study of the Lowell School curriculum pilot. *Environmental Education Research, 26*(4), 511–531. https://doi.org/10.1080/13504622.2019.1607258
Teed, R. E., & Franco, S. (2014). Increasing teachers' confidence and pedagogical content knowledge through a workshop and follow-up program on climate change. *Journal of Geoscience Education, 62*(4), 587–597.
The Seal of Climate Literacy. (n.d.). *The seal of climate literacy*. Retrieved April 23, 2026, from https://www.sealofclimateliteracy.org/seal
UNESCO. (2017). *Education for sustainable development goals: Learning objectives*. UNESCO.
UNESCO. (2021). *Learn for our planet: A global review of how environmental issues are integrated in education*. UNESCO.
United Nations. (2015). *Transforming our world: The 2030 agenda for sustainable development (A/RES/70/1)*.
Wiśniewski, J., & Zahorska, M. (2020). Reforming education in Poland. In F. M. Reimers (Ed.), *Audacious education purposes: How governments transform the goals of education systems* (pp. 181–208). Springer. https://doi.org/10.1007/978-3-030-41882-3_7
World Bank. (2018). *World development report 2018: Learning to realize education's promise*. World Bank.
Zala-Mezö, E., Bormann, I., Strauss, N.-C., & Müller-Kuhn, D. (2019). Distributed leadership practice in Swiss "Eco- Schools" and its influence on school improvement. *Leadership and Policy in Schools*. Advance online publication. https://doi.org/10.1080/15700763.2019.1631855

Open Access This chapter is licensed under the terms of the Creative Commons Attribution-NonCommercial-NoDerivatives 4.0 International License (http://creativecommons.org/licenses/by-nc-nd/4.0/), which permits any noncommercial use, sharing, distribution and reproduction in any medium or format, as long as you give appropriate credit to the original author(s) and the source, provide a link to the Creative Commons license and indicate if you modified the licensed material. You do not have permission under this license to share adapted material derived from this chapter or parts of it.

The images or other third party material in this chapter are included in the chapter's Creative Commons license, unless indicated otherwise in a credit line to the material. If material is not included in the chapter's Creative Commons license and your intended use is not permitted by statutory regulation or exceeds the permitted use, you will need to obtain permission directly from the copyright holder.

Chapter 8
Schools as Anchors of Community Climate Resilience: A Social and Ecological Perspective

Schools are primarily understood as sites of instruction, serving youth within the boundaries of the classroom. However, in practice, schools function as far more than merely educational institutions; in fact, they are social, physical, and cultural infrastructure that influences well beyond enrolled students. Furthermore, they exercise their educational function in interaction with communities and other institutions in the locations where they are placed. The COVID-19 pandemic exemplified this: when schools closed, disruptions occurred in food systems, childcare, health, social services, and more, revealing the larger role schools operate within society (Reimers, 2022). The interaction between schools and climate change is similarly situated and in interaction with communities and place. As already discussed in Chaps. 5 and 6, an effective way to make climate education relevant for students is to situate it in the analysis of climate change in their communities. But the relationship between schools and climate change goes beyond the first and second order effects examined in this book and beyond students learning about climate change, thereby affecting their attitudes and behavior. Schools function as a hub for communities to access credible information and resources in the face of increasing environmental risk, thereby developing their resilience. As Kraft et al. (2025) summarize, the role of schools can expand to be "community hubs that support broader awareness about climate change and care for the planet" for a number of reasons: (1) students influence their parents' and communities' attitudes, (2) most families live close to public schools, (3) schools serve as centers for community, social, and civic functions, and (4) schools provide emergency shelter and care.

This chapter presents an ecological perspective in analysing the relationship of schools to communities as they respond to climate change together. We do this first by examining micro-ecologies of learning and then moving to a macro-level of analysis to examine the relationship between the school and the city. We use a qualitative integrative review and comparative case synthesis to analyze how schools function as community climate infrastructure. First, we will introduce the Reggio

© The Author(s) 2026

F. Reimers and M. Wang-Aghania, *Educating for a Climate Changed Future*,
https://doi.org/10.1007/978-3-032-23256-4_8

Emilia approach as a conceptual lens for understanding schools as embedded socio-ecological systems. Then, we will study empirical conditions that strengthen this role, focusing on schools as emergency coordination hubs, sites of infrastructure adaptation, and community meeting spaces, before synthesizing cross-cutting themes. Lastly, we will shift the analysis to a macro-level perspective, examining Learning Cities as urban-scale learning ecosystems in which education is embedded across municipal systems. Synthesizing this, we conclude with implications for climate change education and policy.

8.1 Reggio Emilia as a Micro-Ecology of Learning

The Reggio Emilia approach emerged after World War II. In the Italian city of Reggio Emilia, families, aligned with local socialist and communist movements, organized collectively under the guidance of Loris Malaguzzi. They wanted to establish new forms of early childhood education that rejected custodial models in favor of democratic and participatory schooling. The city's first municipal preschool opened in 1963, embedding early childhood education within local governance rather than positioning it as a private or peripheral service (New, 2007).

At its core, the Reggio Emilia approach conceptualizes education as a relational and ecological process. Learning is not just the result of the interactions between teachers and children in isolated schools, but emerges through dynamic relationships among children, educators, families, physical environments, and thus, the wider community. Teachers are positioned as co-learners and researchers, while curriculum is formed through extended, inquiry-driven projects that evolve in response to children's interests and exploration rather than a predetermined sequence (New, 2007).

One element that exemplifies schools as anchors for communities is the idea that the physical environment plays a central role in learning. Classrooms, gardens, and public spaces are intentionally designed to provoke curiosity and reflection and function as "niches" that support social growth (Strong-Wilson & Ellis, 2007). Natural environments, in particular, reinforce the importance of outdoor and place-based learning for development (White, 2004). These features have direct relevance for climate change education. By positioning the environment as a co-educator, the Reggio Emilia approach offers a framework for integrating sustainability not only as content but as a lived experience for students. As such, the whole-school approach to sustainability becomes particularly important as climate learning becomes visible beyond the classroom. After all, it is difficult to teach sustainability in settings that materially contradict it. School building and infrastructure themselves become demonstrations of possible futures.

Another element that exemplifies the relationship between schools and communities is the Reggio Emilia principle on shared responsibility and civic participation. Education is understood as a collective endeavor co-produced by schools, families, and municipal institutions; they are a shared civic project for communities (Day,

2007). This orientation situates schools within the fabric of community life rather than apart from it. Public spaces, museums, and civic institutions are treated as extensions of the learning environment, which Malaguzzi called the city's "piazzas"; they are spaces of exchange, collaboration, and learning.

This relational and civic framing has important implications for climate action. The Reggio Emilia model recognizes children as citizens with rights and voices, even in early childhood. This positioning creates pathways through which children's learning can influence household practices and community norms. Research on intergenerational learning supports this idea, showing that climate education can spill beyond classrooms to shape parental attitudes and concern, sometimes particularly among groups less inclined toward climate action (Lawson et al., 2019). Because children's climate perceptions appear less constrained by political ideology, they may serve as effective conduits for building broader social consensus around climate risks and responses.

Therefore, viewed through an ecological lens, Reggio Emilia illustrates how education can function as a micro-ecology that connects learning, environment, family, and civic life. Schools operate as embedded systems that connect individuals, families, institutions, and environments, making them critical anchors of community climate resilience rather than isolated educational sites.

8.2 Conditions that Enhance Schools as Community Infrastructure

While the Reggio Emilia approach provides a conceptual foundation for understanding schools as community infrastructure, there is a body of empirical research and examples that shifts the focus from whether schools serve this role to how their capacity to do so is strengthened. The studies reviewed here do not establish the existence of this role; rather, they examine the conditions that enable schools to function more effectively as community infrastructure for climate adaptation and resilience.

8.2.1 Schools as Emergency Shelters and Coordination Hubs

Across contexts, schools frequently appear in the literature as emergency shelters during floods, storms, and heatwaves (Ardales et al., 2017; López Plazas et al., 2023; Sanz-Mas et al., 2025). However, the evidence suggests that schools' effectiveness as shelters depends less on their physical availability than on the educational and organizational capacities that support coordination.

In northern Thailand, a study of Santikhiri Wittayakom School in Chiang Rai Province examined how universities supported the school's role as a community shelter in a region vulnerable to floods, landslides, and droughts. In partnership with the School of Social Innovation (DRES-ARCID), Kumamoto University, and Hiroshima University, the intervention adopted a holistic disaster risk reduction framework addressing three dimensions of school capacity: institutional systems, physical resilience, and external relations. Rather than prioritizing large-scale infrastructural upgrades, the program focused on training school leaders and teachers in disaster preparedness protocols, emergency response planning, crisis communication, and scenario-based drills. Participants were also trained to pass this knowledge on to others, embedding preparedness within the school's organizational culture (Juwitasari, 2025).

Post-training assessments showed substantial improvements in institutional capacity: understanding of disaster management protocols increased from low baseline scores to near-full confidence, emergency response planning became formalized, and crisis communication systems were modernized. Importantly, these educational gains enabled the school to function as a coordination hub for the wider community. The establishment of an Early Warning System, designed collaboratively with community members and integrating local knowledge with communication technologies, improved timely alerts and collective response. Formal partnerships with local authorities and community organizations were strengthened, and a Community Emergency Response Team (CERT) was introduced, providing structure for search and rescue as well as first aid. These changes increased the school's perceived capacity to serve as a community shelter, even though major physical retrofitting remained limited (Juwitasari, 2025). This case illustrates that schools do not become effective adaptation centers simply by opening their doors during emergencies; rather, education enables schools to activate their physical space as a trusted site for learning, coordination, and collective action.

The experience of Dominica following Tropical Storm Erika (2015) and Hurricane Maria (2017) extends these findings at the scale of an entire education system. In Dominica, schools functioned as emergency shelters during and after both events (Parham, 2022). However, as in the case in Thailand, the literature suggests that the effectiveness of this role depended less on the availability of the school building, but rather the relational and educational capacities surrounding them. A longitudinal qualitative study conducted between 2014 and 2018 examined how Dominica's secondary education system responded to successive climate disasters and the subsequent introduction of national resilience initiatives aligned with the Comprehensive School Safety Framework (CSSF). Drawing on interviews with school leaders, teachers, government officials, NGOs, and disaster risk reduction stakeholders, the study traced how disaster risk reduction education (DRE) was incorporated into schools before and after the disasters.

Prior to Tropical Storm Erika, there was limited evidence of systematic disaster education in schools, and emergency response efforts were largely external to the education sector, led by community disaster teams and humanitarian actors without sustained educational integration (Parham, 2022). Hurricane Maria exposed limitations of this approach and prompted a shift towards the Safer Schools and Smart Schools initiatives as part of a broader effort to "build forward better" toward national climate resilience. These initiatives included nationwide teacher training, the development of school-level disaster management plans, and efforts to integrate multi-hazard disaster risk reduction into curricula. While infrastructural improvements lagged behind immediate needs as schools continued to serve as emergency shelters, the educational interventions produced measurable gains in social resilience. Teachers reported improved understanding of disaster risk, clearer protocols for emergency response, and greater confidence in supporting students during prolonged disruption. At the same time, the literature highlights continual challenges due to uneven teacher preparedness, limited institutional capacity to sustain these reforms, and overreliance on short-term NGO supports, highlighting the importance of sustained educational efforts. Despite this, schools became sites where disaster preparedness knowledge was transmitted beyond students to families and communities, reinforcing their role as trusted coordination hubs during crises. Disaster risk reduction education transformed schools from passive sites of refuge into active nodes of community resilience.

Together, these cases extend the Reggio Emilia conception of schools as community-embedded learning environments, showing how education enables schools to operate as community infrastructure for coordination and collective resilience in the face of climate disruption.

8.2.2 Schools as Infrastructure Adaptation

Schools are also uniquely positioned to serve as more than emergency coordination hubs; they can model adaptive infrastructure that benefits both students and the wider community. Unlike other public buildings, schools are sites of trust: parents, educators, and local residents all have shared investment in ensuring that these spaces are safe and conducive to learning. Moreover, schools make up a significant portion of urban and suburban built environments, meaning interventions at school sites have potential to influence city-wide resilience. As a result, schools can act as living models for climate adaptation, demonstrating strategies that can be scaled across cities. This concept resonates with Reggio Emilia's approach to education: learning environments should be integrated into the life of the community, not isolated, and can serve as sites where students experience and shape the future they hope to live in.

In Barcelona, a pilot program transformed 11 public primary schools into "climate shelters" by integrating green, blue, and grey infrastructure: planting trees, installing shade structure, adding water features, and improving building insulation and ventilation. Although this initiative was done to provide shelters during extreme heat, the schoolyards were opened to the public during non-school hours, extending benefits to the whole community beyond emergencies. The program replaced approximately 1000 m^2 of concrete with natural surfaces, planted 74 trees, generated over 2200 m^2 of shade, and installed 26 new water sources. A mixed-method evaluation assessed environmental conditions, health and well-being outcomes for students and teachers, and community use during summer months. The findings suggest that schools can operate as neighborhood-scale cooling infrastructure while simultaneously supporting social interaction. The project positioned schools not only as sites of protection, but as spaces where climate adaptation is experienced (Sanz-Mas et al., 2024).

In Tainan, Taiwan, the Ministry of Education implemented a holistic water management system at Taikang Elementary School, including underground water storage, reconnected circulation for stormwater, and natural infiltration irrigation. These interventions allowed the campus to withstand heavy rainfall events without flooding and supplied over 90% of the irrigation water needed annually. The system also improved thermal comfort via green and water areas that cooled naturally ventilated classrooms. Community involvement was central to the project, extending the Reggio Emilia's philosophy that education sites are co-developed by the community. In addition, the schools adopted place-based curricula aligned with their campus renovations. For example, the water management system became the foundation for a vertically integrated curriculum with teachers across science and social studies collaborating to design the scope and sequence (Li et al., 2020).

Szombathely, Hungary demonstrated that rooftop greening on a single school can reduce cooling energy demand by over 22% during peak summer months and improve thermal comfort for surrounding public spaces, emphasizing the role of schools as durable infrastructure for communities (Li et al., 2025). In addition, this study shows that when scaled, such interventions could create interconnected green networks across urban areas. Schools were identified as key public sites to coordinate across multiple public buildings to not only be energy-saving interventions, but a tool for urban-scale green networks for social interaction, redefining the social and environmental functionality of public buildings.

These cases collectively illustrate the concept of schools as microcosms of resilient communities: infrastructure interventions not only protect students but also provide living examples of sustainable design that can be replicated and scaled. Schools' physical presence, combined with their social centrality, enables them to serve as both practical and symbolic demonstrations of adaptive futures.

In the United States, initiatives such as the Aspen Institute's K–12 Climate Action framework further underscore the transformative potential of schools. Serving nearly one in six Americans, K-12, their motivation statement highlights that schools can simultaneously reduce carbon emissions, increase resilience, provide climate education, and advance equity. The framework emphasizes four pillars, mitigation, adaptation, education, and equity, ensuring that schools are not only prepared for climate impacts but are also active sites for fostering sustainable practices and empowering youth to shape the future (Aspen Institute, n.d.).

By integrating climate-resilient infrastructure with community engagement and learning opportunities, schools can shift from reactive spaces to proactive exemplars. As Reggio Emilia reminds us, these environments should be celebrated as living, participatory spaces, not only responding to disasters, but shaping how communities envision and learn about a sustainable and resilient future.

8.2.3 *Schools as Community Meeting Spaces*

Schools not only function as climate-resilient infrastructure beyond moments of disasters, but they also operate as hubs of interaction, social connection, and capacity-building. Evidence from multiple contexts suggests that schools become effective resilience hubs not primarily because they provide shelter or services, but because they are embedded in everyday community life. In Dominica, for example, initiatives such as Child Friendly Spaces and the Return to Happiness (RTH) programme enabled schools to offer structured opportunities for students to process traumatic experiences and regain a sense of safety following Hurricane Maria. While educators acknowledged that pre-disaster instruction on hazards had been insufficient, the post-Maria training substantially improved their ability to respond to students' emotional needs (Parham, 2022). In this way, schools functioned not only as places of learning, but as trusted community spaces where social reconnection and collective healing could take place.

In Veracruz, Mexico, high schools similarly served as platforms for resilience-building after Hurricane Karl, but in a different manner. Through participatory research, workshops, and training activities, schools became convening spaces where students, teachers, and local actors co-produced knowledge about risk and preparedness. High school students emerged as community change agents, organizing prevention brigades, facilitating community assemblies, and engaging local authorities (Maldonado-González et al., 2019). These processes strengthened communication networks and fostered shared responsibility for risk management, illustrating how schools can anchor not only physical gathering but also civic engagement and political capacity at the community level.

Social resilience outcomes can also emerge less through explicit disaster programming and more through the cultivation of everyday social cohesion. In Barcelona, the "Climate Shelters in Schools" initiative transformed schoolyards in publicly accessible cooling and gathering spaces through vegetation, shade, water infrastructure, and seating. By opening schoolyards to the public outside school hours, the city repositioned schools as neighborhood-scale climate infrastructure that simultaneously supported recreation and social interaction. Empirical studies of these transformed schoolyards show that they function as everyday meeting spaces for families. These benefits extended beyond climate protection alone: shaded seating and play areas encouraged intergenerational interaction and community-bonding (López Plazas et al., 2023). Evidence from Barcelona also highlights the importance of intentional design and programming. While schoolyards attracted high levels of use among children and adolescents, particularly boys, lower participation among girls and older adults underscored the need for inclusive programming and community engagement. In other words, schools do not automatically become inclusive community spaces through physical transformation alone. Their effectiveness as social infrastructure depends on communication strategies and opportunities for community participation that actively invite diverse users into the space.

Similarly, in the United States, renovated schoolyards that integrate natural elements have been shown to promote sustained physical activity, positive social interaction, and perceptions of safety among children and caregivers, with effects maintained years after intervention. These spaces are used throughout the day and across age groups, positioning school grounds as accessible public spaces in neighborhoods often lacking green infrastructure (Bates et al., 2018). An interesting conclusion from the study highlighted that when communities are involved in the design of these spaces, schoolyards become shared civic assets rather than perceived sources of risk. Therefore, becoming more of a community space for intergenerational gathering, supporting neighborhood cohesion. Initiatives supported by organizations such as Green Schoolyards America emphasize this systems-level approach, framing school grounds not only as green infrastructure for cities but as vibrant neighborhood centers where students and community members actively shape and care for a shared public space (Fine, 2025).

These cases demonstrate that schools function as community meeting spaces not as an ancillary benefit of adaptation, but as a core mechanism through which resilience is built. By hosting everyday social interaction, facilitating dialogue, and embedding learning within shared spaces, schools cultivate the relational foundations on which effective climate response depends. In this sense, schools extend beyond their institutional role to become civic anchors, linking education, public space, and community life into a durable infrastructure for resilience.

8.2.4 Cross-Cutting Themes: Schools as Community Climate Infrastructure

Across the cases reviewed, schools function as effective community climate infrastructure because they integrate social, educational, and physical capacities within trusted institutions. Rather than operating solely as sites of service delivery or emergency response, schools act as relational infrastructure knowledge and collective action converge.

First, community capacity building is the most consistently documented outcome. In Thailand, Dominica, and Mexico, schools served as training and coordination centers for disaster preparedness, emergency response, and risk communication (Juwitasari, 2025; Maldonado-González et al., 2019; Parham, 2022). These capacities were not limited to students but extended outward to families. Whether through Community Emergency Response Teams, youth-led brigades, or participatory risk assessments, schools anchored the development of local skills and organizational readiness. Importantly, the literature shows that these capacities were strongest where training was embedded in ongoing educational practices rather than delivered as one-off interventions.

Second, social cohesion and connectivity emerged as a critical, though often less formally measured, dimension of resilience. Schools provided spaces where communities gathered not only during crises but also in everyday life through open schoolyards, community workshops, psychosocial support programs, and participatory infrastructure projects. In Barcelona and the United States, climate-adapted schoolyards functioned as everyday meeting spaces that encouraged informal and intergenerational contact, leading to a sense of shared ownership (Bates et al., 2018; Fine, 2025; Sanz-Mas et al., 2025). In Mexico and Taiwan, collaborative projects strengthened communication networks and fostered shared responsibility for managing environmental risk (Li et al., 2020; Maldonado-González et al., 2019). These relational effects formed the social cohesion that enables more effective collective response during disruptions.

Third, knowledge transfer across generations and sectors was a defining feature of schools as community infrastructure. Students frequently acted as intermediaries between formal knowledge and household or community practices, whether through disaster risk education or learning about their schools' infrastructure improvements. This aligns with evidence that schools can amplify learning beyond their walls, particularly when curricula are place-based and linked to lived experience. Across cases, the most durable outcomes occurred where physical infrastructure interventions, such as water systems or green roofs, were directly integrated into teaching and learning.

Fourth, equity and accessibility shaped both the reach and effectiveness of school-based interventions. While schools are widely distributed and generally accessible, the literature highlights uneven participation by age, gender, and social group, particularly in open schoolyard and public space initiatives. These findings suggest that schools do not automatically function as inclusive community infrastructure; their effectiveness depends on intentional design and programming that actively engage diverse users. Where such efforts were made, schools helped mitigate inequalities in access to green space, cooling infrastructure, and learning opportunities.

Finally, the review highlights the dual role of schools as both physical and social infrastructure. Quantitative evidence demonstrates that school-based interventions can measurably improve flood protection, thermal comfort, energy efficiency, and service continuity (Juwitasari, 2025; Li et al., 2020, 2025; Sanz-Mas et al., 2024). At the same time, these physical improvements were most impactful when coupled with social processes such as training and learning. Schools thus operate as microcosms of resilient communities where adaptation is not only implemented, but made visible, learnable, and transferable.

This framing builds directly on the Reggio Emilia conception of schools as embedded, relational ecologies and points toward a larger question of scale. If schools can function as micro-ecologies of resilience within neighborhoods, how might education systems operate at the level of cities themselves? The following section addresses this question by turning to Learning Cities as macro-ecologies of learning and climate action.

8.3 Learning Cities as Macro-Ecologies of Learning

While the Reggio Emilia approach illustrates how schools and local communities constitute a micro-ecology to respond to climate change, the concept of Learning Cities extends this ecological perspective to urban systems and governance. Learning Cities conceptualize education not as the responsibility of a single sector, but as a city-wide ecosystem involving schools, workplaces, cultural institutions, civil society organizations, and municipal authorities. Education becomes both a driver and an outcome of broader social, economic, and environmental development.

The UNESCO Global Network of Learning Cities, launched in 2013, is a network of cities mobilizing resources across sectors to support lifelong learning, social inclusion, economic development, cultural vitality, and sustainability (UNESCO, n.d.). Rather than focusing on formal schooling, Learning Cities emphasizes learning in communities to achieve all 17 Sustainable Development Goals (SDGs).

The relationship between education and other sectors became visible in Learning Cities during the COVID-19 Pandemic. As cities emerged as the center of this crisis, municipal responses depended on cross-sectoral collaboration. Libraries, museums, schools, community organizations, health services, and digital platforms were mobilized to maintain educational continuity and disseminate public health information (UNESCO, 2020). These experiences underscored that education served as a connective tissue linking health, social protection, communication, and civic engagement.

Insights were synthesized in the report *From Emergency to Resilience: Building Health and Resilient Cities through Learning*, which positioned education in Learning Cities as central not only to crisis response but to sustained resilience-building after the pandemic (UNESCO Institute for Lifelong Learning, 2022). A cluster of themes focused on how education is key for health and well-being outcomes; another theme centered around sustainability. Regional Centres of Expertise (RCE) were presented as examples. In Denmark, RCE created apprenticeship programmes for students to build career-related skills through a green lens: transforming regular construction into sustainable methods. Education emerged as the mechanism through which communities developed the knowledge, skills, and relationships necessary to respond to complex challenges, including climate change (UNESCO Institute for Lifelong Learning, 2022).

The relationship between health, sustainability, and education is also emphasized in the report. The report examines how learning initiatives support physical activity, healthy diets, environmental stewardship, access to green spaces, and healthier working environments, highlighting education's role in shaping everyday conditions that influence both human and ecological well-being. For example, leadership from Ybycuí, Paraguay emphasized the importance of building social structures that advance sustainable development across society, combining environmental protection with cultural and educational initiatives such as communal libraries, intergenerational festivals, and arts education. These efforts demonstrate how education, culture, and environmental planning can operate together as mutually reinforcing components of local sustainability strategies (UNESCO Institute for Lifelong Learning, 2022). The discussions reflected in the report underscored that climate resilience is not solely a matter of environmental policy, but is deeply intertwined with public health and social sustainability.

UNESCO's Institute for Lifelong Learning further reinforces this cross-sectoral relationship with education at its core through its recent guidance, *Greening communities guidance: lifelong learning for climate and sustainability action,* which outlines how cities can mobilize education as a strategy for climate solutions for all community members (UNESCO, 2025). Rather than positioning education as the responsibility of a single sector, the guide explicitly lists out several stakeholders that can support greening communities (government, civil society, education and academia, change agents, private sector). This understores education's role not as a discrete intervention, but as an integrative mechanism across all sectors. For example, in Bogotá, the Environmental School Project brings together the District Secretary of Education, Mayor's Office of Bogota, Colombian Academy of Sciences and community partners to embed sustainability learning across formal education and city systems. Through partnerships with environmental groups and the manufacturing sector, students engage with local environmental challenges, while educators participate in blended professional learning on climate change developed in collaboration with the national scientific community. By 2023, this initiative has supported 300 educators, demonstrating the principles associated with Reggio Emilia (learning from and through the environment as the third teacher) can be extended through cross-sectoral partnerships. This is an example of how other sectors can collaborate with the education sector to provide these learning opportunities.

In other cases, education can become a core strategy for other sectors not traditionally framed as educational. In communities across Nepal and Uganda, intergenerational dialogue has been used as a learning-based approach to locally led climate action through the respective local chapter of HelpAge. Structured conversations brought younger and older community members together to share experiences of environmental change, discuss coping strategies, and identify pathways for mitigation and adaptation. These dialogues were facilitated in local languages and grounded in respect for traditional knowledge alongside contemporary sources of information. The outcomes extended beyond dialogue itself: local partner organizations received small grants to support follow-on climate actions, including public forums, food security initiatives, and environmental restoration efforts. These cases highlight how learning can reduce social fragmentation while strengthening collective capacity to respond to climate risks. In addition, they exemplify how learning functions as the enabling component for action in other sectors.

Similarly, UNESCO (2025) shows that learning can occur in environments beyond schools. Climate-relevant learning is shown to take place across agricultural extension offices, natural and community spaces, municipal institutions, workplaces, transit systems, digital platforms, etc. (UNESCO, 2025). In the agricultural sector, for example, the UN's Food and Agricultural Organization's Global Soil Doctors Programme operates as a farmer-to-farmer learning model, building practical knowledge around sustainable soil management while supporting national and regional climate and food security goals (UNESCO, 2025). By strengthening farmers' decision-making and ecological practices, the programme illustrates how education embedded within systems can contribute directly to environmental resilience.

Similar dynamics are evident in the tourism sector. In Wujing Town, China, environmental learning is integrated into a public walking route that combines cultural heritage, ecological observation, and scientific inquiry. Led by the Department of Environmental Science at East China Normal University (ECNU), children engage with local flood management infrastructure, river ecology, and water quality through guided observation and hands-on data collection. Activities such as river water sampling and ecological analysis transform public space into a learning environment, fostering environmental awareness while strengthening young people's connection to place (UNESCO, 2025).

These examples show that education functions as a central throughline for climate action because it is inherently cross-sectoral. Learning Cities represent a macro-ecological extension of the Reggio Emilia philosophy: where Reggio Emilia demonstrates how schools can anchor learning within communities, Learning Cities show how education can be embedded within municipal systems themselves. In doing so, education links health, culture, environment, and governance, serving as a key mechanism through which cities build the social capacity required to respond to climate change throughout several sectors.

8.4 Implications for Climate Change Education and Policy

This chapter suggests that climate change education cannot be understood as a curricular intervention alone, but as a catalyst that enhances the effectiveness of climate adaptation and mitigation actions.

The Reggio Emilia approach and the Learning Cities framework both illustrate how schools can be designed and governed as ecological and civic institutions. In these models, learning is embedded in place, shaped through interaction with the environment, and sustained through collaboration among educators, families, public agencies, and civil society. This expands the scope of climate education from individual knowledge acquisition to the cultivation of social, emotional, and political capacities necessary for long-term resilience.

As a result, several implications for climate education policy follow:

First, schools should be part of climate adaptation plans, disaster risk reduction strategies, heat action responses, and other community resilience plans not only as shelter, but as coordination and learning sites. Schools should be treated as nodes within local climate ecosystems, rather than as isolated academic institutions. Their physical infrastructure and everyday presence position them to support adaptation, preparedness, and recovery in ways that other public institutions cannot. Climate resilience planning that overlooks schools risks missing one of the most trusted and widely distributed forms of community infrastructure.

Second, climate adaptation and resilience investments should include dedicated education components in order to amplify and scale impact. Physical interventions such as cooling infrastructure, green schoolyards, or disaster management achieve more durable and equitable outcomes when communities understand their purpose, participate in their use, and develop the skills needed to maintain and adapt them over time. Education is one of the primary mechanisms through which such capacity is built.

Third, climate education policies should not only focus on educating students, but also engaging families and communities. On one hand, schools offer uniquely scalable pathways for climate literacy, given the number of students they reach. On the other hand, schools are also positioned to engage families and communities not only through students' influences, but also because schools are shared learning spaces. In addition, schools represent participatory processes in which future pathways are actively negotiated and shaped. Through shared use of school spaces, place-based learning, and visible representations of climate adaptation embedded in everyday environments, schools extend education into the public realm.

Recognizing and supporting this role reframes climate education as a form of public infrastructure for responding to climate change at scale.

References

Ardales, G., Jr., Espaldon, M. V., Lasco, R., Quimbo, M. A., & Zamora, O. (2017). Adaptive capacity index of public schools in the municipalities of bay and los baños, Laguna, Philippines. *Journal of Environmental Science and Management, 20*(1), 71. https://doi.org/10.47125/jesam/2017_1/08

Aspen Institute. (n.d.). *K12 climate action.* This is planet ed. Retrieved January 30, 2026, from https://www.thisisplaneted.org/initiatives/k12-climate-action

Bates, C. R., Bohnert, A. M., & Gerstein, D. E. (2018). Green schoolyards in low-income urban neighborhoods: Natural spaces for positive youth development outcomes. *Frontiers in Psychology, 9.* https://doi.org/10.3389/fpsyg.2018.00805

Fine, E. (2025, June 26). Early impacts and positive results from schoolyard forest leadership institute. https://www.greenschoolyards.org/blog/2025/6/23/leadership-institute-early-impacts

Juwitasari, R. & Disaster Resilience and Environmental Sustainability (DRES) Program, Asian Research Center for International Development (ARCID), School of Social Innovation, Mae Fah Luang University, Chiang Rai 51700, Chiang Rai Province, Thailand. (2025). Strengthening resilience in schools and communities: The strategic role of universities in climate crisis management. *Journal of Humanities and Social Sciences Research, 7*(S), 106–116. https://doi.org/10.37534/bp.jhssr.2025.v7.nS.id1299.p106-116

Lawson, D. F., Stevenson, K. T., Peterson, M. N., Carrier, S. J., Strnad, L. R., & Seekamp, E. (2019). Children can foster climate change concern among their parents. *Nature Climate Change, 9*(6), 458–462. https://doi.org/10.1038/s41558-019-0463-3

Li, X., Zhao, T., & Kósa, B. (2025). Beyond energy savings: The role of green roofs in public buildings for climate resilience and peri-urban renewal—Insights from Hungary. *Inżynieria Mineralna, 2*(2). https://doi.org/10.29227/IM-2025-02-02-097

Li, Y. Y., Wei, C. A., Jing, S. R., Lin, K. L., & Li, C. T. (2020). Integrated renovation practices on water resilient infrastructure in Tainan Taikang elementary school in Taiwan. *IOP Conference*

Series: Earth and Environmental Science, 588(5), 052007. https://doi.org/10.1088/1755-1315/588/5/052007

López Plazas, F., Crespo Sánchez, E., Llorca Pérez, R., & Santacana Albanilla, E. (2023). Schools as climate shelters: Design, implementation and monitoring methodology based on the Barcelona experience. *Journal of Cleaner Production, 432*, 139588. https://doi.org/10.1016/j.jclepro.2023.139588

Maldonado-González, A. L., Cruz-Sánchez, G. E., Bello-Benavides, L. O., & González-Gaudiano, E. J. (2019). Shared commitments towards social resilience in populations vulnerable to extreme weather conditions. *Southern African. Journal of Environmental Education, 35*(1). https://doi.org/10.4314/sajee.v35i1.3

Matthew, A. K., Malik, S., & Falken, G. (2025). *Education and climate change: Synthesizing the evidence to guide future research.* (EdWorkingPaper: 25-1238). Annenberg Institute at Brown University. https://doi.org/10.26300/tgea-5e35

New, R. S. (2007). Reggio Emilia as cultural activity theory in practice. *Theory Into Practice, 46*(1), 5–13. http://www.jstor.org/stable/40071628

Parham, M. (2022). Returning to normal? 'Building back better' in the Dominican education system after tropical storm Erika and Hurricane Maria. *Disasters, 46*(S1). https://doi.org/10.1111/disa.12536

Reimers, F. M. (2022). Learning from a pandemic. The impact of covid-19 on education around the world. In F. M. Reimers (Ed.), *Primary and secondary education during Covid-19: Disruptions to educational opportunity during a pandemic* (pp. 1–37). Springer International Publishing. https://doi.org/10.1007/978-3-030-81500-4_1

Sanz-Mas, M., Continente, X., Marí-Dell'Olmo, M., & López, M. J. (2025). Community use and perceptions of climate shelters in schoolyards in Barcelona. *International Journal of Public Health, 70*, 1608083. https://doi.org/10.3389/ijph.2025.1608083

Sanz-Mas, M., Ubalde-López, M., Borràs, S., Brugueras, S., Continente, X., Daher, C., Marí-Dell'Olmo, M., & López, M. J. (2024). Adapting schools to climate change with green, blue, and grey measures in Barcelona: Study protocol of a mixed-method evaluation. *Journal of Urban Health, 101*(1), 141–154. https://doi.org/10.1007/s11524-023-00814-y

Strong-Wilson, T., & Ellis, J. (2007). Children and place: Reggio Emilia's environment as third teacher. *Theory Into Practice, 46*(1), 40–47. http://www.jstor.org/stable/40071632

UNESCO. (n.d.). *UNESCO global network of learning cities.* UNESCO Institute for Lifelong Learning. Retrieved January 30, 2026, from https://www.uil.unesco.org/en/learning-cities

UNESCO. (2020). *How cities are utilizing the power of non-formal and informal learning to respond to the COVID-19 crisis.* UNESCO Institute for Lifelong Learning. https://unesdoc.unesco.org/ark:/48223/pf0000374148

UNESCO. (2025). *Greening communities guidance: Lifelong learning for climate and sustainability action.* UNESCO Institute for Lifelong Learning. https://unesdoc.unesco.org/ark:/48223/pf0000396434

UNESCO Institute for Lifelong Learning. (2022). *From emergency to resilience: Building healthy and resilient cities through learning: Fifth international conference on learning cities, Yeonsu, Republic of Korea, 2022.* UNESCO. https://unesdoc.unesco.org/ark:/48223/pf0000380657

White, R. (2004). Interaction with nature during the middle years: Its importance in children's development and nature's future. Retrieved October 28, 2004, from http://whitehutchinson.com/cgibin/printer=/children/articles/nature.shtml

Open Access This chapter is licensed under the terms of the Creative Commons Attribution-NonCommercial-NoDerivatives 4.0 International License (http://creativecommons.org/licenses/by-nc-nd/4.0/), which permits any noncommercial use, sharing, distribution and reproduction in any medium or format, as long as you give appropriate credit to the original author(s) and the source, provide a link to the Creative Commons license and indicate if you modified the licensed material. You do not have permission under this license to share adapted material derived from this chapter or parts of it.

The images or other third party material in this chapter are included in the chapter's Creative Commons license, unless indicated otherwise in a credit line to the material. If material is not included in the chapter's Creative Commons license and your intended use is not permitted by statutory regulation or exceeds the permitted use, you will need to obtain permission directly from the copyright holder.

Chapter 9
Conclusion: Reorienting Education in a Warming World

This book began by posing three sets of questions.

- Conceptualization. How should we understand the relationship between education and climate change? In what ways are they interdependent systems? How do different narratives, climate literacy, education for climate action, green-skills agendas, education for sustainable development, critical and decolonial approaches, frame education's role?
- Policy and Practice. How do these aims translate into national systems, informal networks, schools, classrooms, and ultimately, student outcomes? In other words, what is being done, and what evidence do we have of effects on learning, behavior, policy and climate outcomes?
- Gaps and opportunities. Where are the mismatches between ambitions and reality? Who and what are neglected? What opportunities exist to align education systems more closely with the demands of a rapidly warming world?

This concluding chapter synthesizes what the book has shown in response to these questions and sketches what it would mean to act on that knowledge.

9.1 Conceptualization: The Education–Climate Nexus

9.1.1 Two Complex Systems in Interaction

The book has argued that "education" and "climate" are each a complex, adaptive system, and they are mutually constitutive. The climate system consists of interacting subsystems, atmosphere, hydrosphere, cryosphere, land and biosphere, linked through non-linear feedbacks, thresholds and tipping points. Human activities, especially greenhouse gas emissions and land-use change, are now a dominant contributor.

© The Author(s) 2026
F. Reimers and M. Wang-Aghania, *Educating for a Climate Changed Future*,
https://doi.org/10.1007/978-3-032-23256-4_9

The education system comprises institutions, policies, curricula, assessments, funding flows, professionals, organization, leadership, governance and norms. It, too, exhibits emergent properties: what students learn results from the interactions among these elements, not simply from what is written in policy.

From this system's point of view, the relationship between education and climate change is two-way. Climate change affects education first-order, by disrupting schools, health and livelihoods. Education affects climate second-order, by shaping knowledge, values, behaviors, institutions and technologies that, over time, influence emissions, adaptation choices and political will.

These interactions play out in various timescales and are mediated by feedback loops and by larger governance, economic and cultural arrangements. For example, a heatwave that lowers test scores may trigger investments in school cooling or shifts in schedules; a cohort of climate-literate graduates may, decades later, change voting patterns or professional practices.

9.1.2 Timescales and Life-Course

These relationships need to be understood with attention to the timescales at which they operate. While there may be synchronous effects, for instance students missing classes because of floods, or perform poorly on a test because of extreme heat, there are deep effects in this relationship that are not synchronous. Climate change unfolds over decades and centuries; many of its near-term impacts are already "committed" by past emissions. Educational effects likewise accrue over the life-course: what children learn now may shape their decisions, livelihoods and political behavior long after they leave school, and the way they live will impact climate over decades into the future.

This temporal understanding is critical. Schools today are dealing with climate impacts rooted in decisions made long ago, while educating students whose most consequential climate actions may take place in futures that neither they nor their teachers will inhabit in full. Evaluating climate-related education is therefore inherently difficult: the outcomes that matter most, lifetime emissions, adaptive capacity, civic engagement, lie beyond typical research and policy horizons.

Recognizing this does not make short-term outcomes irrelevant; it does underscore the need for life-course and intergenerational perspectives in both research and policy.

9.1.3 First- and Second-Order Effects

We have offered the distinction between first-order and second-order as a practical organizing frame:

First-order effects are the ways climate change directly and indirectly disrupts education: heat stress and air pollution that damage cognition; floods and storms that destroy schools and close them for weeks; droughts and crop failure that pull children out of school; displacement that uproots learners and overburdens host systems.
Second-order effects are the educational responses and contributions: curricular reforms; climate-focused teacher education; green and resilient school infrastructure; greening of school operations; development of green skills; and efforts to reposition schools as community hubs for climate awareness and action.This lens helped keep in view that adaptation and mitigation are not add-ons but cut across infrastructure, operations, teaching, student support and community roles.

9.1.4 Multiple Narratives of Climate Change Education

In Chap. 2 we examined five co-existing narratives about education's climate role:

- Climate literacy: education should primarily develop robust understanding of climate science and its implications.
- Education for climate action: knowledge must be linked to individual and collective action, changing everyday practices, social norms and participation in climate politics.
- Education for the green economy: systems must prepare learners with the skills needed for low-carbon, climate-resilient economies, from technical green skills to transversal "green life skills" and transformative capacities.
- Education for sustainable development (ESD): climate change should be addressed within a holistic agenda that links environmental integrity, social justice and economic viability, with emphasis on cross-curricular integration and transformative pedagogy.
- Critical and decolonial climate education: climate change must be understood in relation to histories of colonialism, capitalism and structural inequality; Indigenous and local knowledge systems, gender justice and political agency are central, not peripheral.

Policies and programs rarely adopt any of these narratives in pure form; instead, they braid elements of several, often without naming the tensions. For example, a national curriculum might combine scientific literacy and green-skills language, while a school-based program weaves together climate science, youth activism and local Indigenous knowledge.

Rather than seeking to integrate these narratives into a single model, the book has argued for pluralism with reflexivity, making explicit which narratives are foregrounded, with what assumptions, and asking which combinations best fit contexts and purposes.

9.1.5 *Global Governance and "Green Education"*

Conceptually, the book has also located climate change education within a rapidly evolving architecture of global education governance.

From early environmental education declarations (Belgrade, Tbilisi) through Agenda 21 and the UN Decade of Education for Sustainable Development, to the ESD for 2030 roadmap and the Greening Education Partnership, frameworks from multilateral organizations have progressively:

- Tied environmental and climate concerns to the right to education.
- Defined broad aims and dimensions of "green education" (governance, transformative teaching and assessment, lifelong learning ecosystems, sustainable infrastructure).
- Promoted Action for Climate Empowerment under the UNFCCC, making education a formal component of climate agreements; and
- Articulated standards and tools (e.g., global learning objectives, green-school quality standards, curriculum guidance).

At the same time, comparative research shows that these norms are translated unevenly into national policies, curricula and classrooms, and can be resisted, reshaped or diluted by local politics and markets. Understanding the education–climate nexus thus requires attention not only to ideas and standards, but to how they travel through and interact with national and local systems.

9.2 Policy and practice: What Systems, Schools and Communities Are Doing

Turning to policy and practice, the book has assembled evidence at multiple levels: global reviews, national case studies, teacher and school networks, classroom interventions, student assessments and cross-sector collaborations.

9.2.1 *First-Order Impacts and Adaptation*

Empirical syntheses show that climate change is already a systemic driver of educational disruption, not an occasional shock. Heat lowers test scores and slows learning, with the strongest effects where buildings lack cooling and ventilation and among younger, poorer and marginalized students. Disasters, typhoons, floods, droughts, wildfires, damage or destroy thousands of schools each year, close them for extended periods, and divert buildings into use as shelters. Slow-onset changes erode livelihoods, increase climate-related disease and exacerbate poverty, all of which contribute to absenteeism, dropout and lowered attainment. Displacement creates new barriers to education for internally displaced and migrant learners, and strains host schools.

The book has also documented emerging adaptation strategies: risk-informed school-building standards and retrofits; relocation of schools from high-risk zones; hybrid and alternative learning modalities for periods of closure; heat-wave protocols and schedule adjustments; and early-warning and continuity plans that tie education decisions to hazard information.

Evidence suggests that such measures can shorten closures and protect learning, but they remain piecemeal and underfunded, especially in low-income and fragile contexts. Education plans and climate strategies rarely integrate adaptation for education at scale.

9.2.2 Policy Ambition Meets System Fragmentation

This book shows how national governments are operationalizing climate-ready education systems as Chap. 3 focuses on policy instruments, governance arrangements, and implementation mechanisms. Governments increasingly recognize climate education in law, policy, and curriculum. Yet this recognition rarely translates into coherent systems. Curriculum presence often does not translate into classroom practice, climate content is uneven and largely cognitive, teacher capacity remains a central bottleneck, and monitoring infrastructure is weak. Different pathways (legalistic, curriculum-driven, green-school programs) each have strengths and vulnerabilities, but none succeed without aligned professional learning and assessment. System coherence, not absence of policy, is the dominant challenge.

Where system coherence falters, Chap. 4 shows that educator networks fill critical gaps. School- and district-based PLCs, international teacher networks, digital communities, and researcher partnerships function as protected spaces for experimentation, professional learning, and diffusion of practice. These networks cultivate trust, agency, and shared norms, enabling climate education to scale even when formal systems are fragmented. However, participation is uneven, and networks cannot replace system reform; they are complements that make systemic transformation more likely.

9.2.3 Practices in Schools and Classrooms

What could climate change education look like when enacted classrooms, moving from policy texts to lived educational experiences? Chap. 5 provides analytical windows through global reviews of research and in-depth cases, not meant as exemplars to be copied, but revealing which combinations of narratives are viable in practice.

At the micro-level, the book has reviewed a wide range of instructional approaches:

- inquiry-based climate science lessons using real data or models.
- place-based projects linking local hazards, ecosystems and livelihoods.
- humanities-based curricula using fiction, narratives and the arts to explore climate stories and emotions.
- justice-centered projects that situate climate in broader struggles over land, work, gender and race.
- youth-led programs that invert traditional hierarchies (students teaching teachers, or co-leading research and advocacy).
- land-based and Indigenous programs where climate learning is embedded in stewardship, language and cultural practice.

Systematic reviews and case studies converge on several design principles: personal relevance, active participation, opportunities for dialogue and deliberation, attention to misconceptions, and genuine action projects all contribute to deeper understanding and more engaged attitudes. The most powerful programs braid cognitive, socio-emotional and action dimensions rather than treating climate as a purely scientific topic. Further, deep integration, where climate and sustainability are organizing principles of learning, school identity, and community engagement, produces multi-year projects, visible feedback loops, and meaningful student agency.

Teacher capacity is shown to be system-produced: it grows when teachers are supported through networks, partnerships, and permission to experiment.

9.2.4 *What Students Know, Feel and Do*

Yet innovation remains fragile and uneven without structural support. As a result, Chap. 6 highlights resulting student learning, attitudes, and behavioral outcomes. Empirical evidence highlights the paradox that high concern and enthusiasm among students coexist with fragmented scientific understanding, persistent misconceptions, and limited engagement in high-impact actions. For example, many students can name basic causes and impacts of climate change, but understanding of core mechanisms (the enhanced greenhouse effect, role of different gases, feedbacks) is often partial and fragile. Misconceptions are widespread and persistent: confusion between climate and weather; conflation of climate change with ozone depletion or generic "pollution"; misunderstandings of the time scales involved. Furthermore, most young people express concern about climate change and support for environmental protection. In many countries, they see climate action as a civic responsibility. Yet there is a pronounced attitude–action gap. Students gravitate toward low-cost, low-disruption actions (recycling, switching off lights, planting trees), and tend to underestimate the importance of high-impact actions (e.g., changes in transport, diet or consumption). Many perceive climate change as both urgent and psychologically distant, global, future, or someone else's responsibility. Feelings of powerlessness and anxiety are common.

The attitude–action gap is not a motivation problem but a learning design problem. Effective climate education integrates scientific foundations, personal relevance, and collective, real-world action. Educational interventions can and do improve knowledge and, to a lesser extent, self-efficacy and behavioral intentions. But robust evidence on long-term behavioral and emissions outcomes remains limited. This underlines the need to treat claims about education's climate impact as hypotheses to be tested, not assumptions.

9.3 Gaps Between Ambition and Reality, and Opportunities Ahead

9.3.1 Frameworks to Understand Education Systems

Many gaps in climate education persist not due to lack of frameworks, evidence, or examples, but because systems do not see, reward, or protect climate learning. In Chap. 7, we argue that curriculum, assessment, professional development, leadership, and governance often send conflicting signals. The book introduces a systems lens that identifies "low-climate-learning traps": stable equilibria, misaligned incentives, weak information, and fragmented authority, that reproduce thin, cognitive, and fragile climate education. Multidimensional reform requires cultural, psychological, professional, institutional, and political alignment; futures thinking; and iterative stewardship rather than one-off interventions.

Finally, Chap. 8 expands the unit of analysis beyond classrooms and systems to schools as physical, social, and educational infrastructure. Schools are multifunctional sites for learning, trust-building, and civic engagement. When learning is embedded in place, connected to community, and supported across municipal systems, climate education multiplies the impact of public infrastructure, strengthens resilience, and fosters intergenerational knowledge transfer. This reframing positions schools as central actors in adaptation and mitigation, not merely sites of curriculum delivery.

9.3.2 Gaps

Across the book, several consistent mismatches emerge:

- **Thin integration and misalignment**: Global and national rhetoric increasingly presents education as central to climate solutions, but climate change is still

absent from many curricula, and where present, treated superficially. Climate-related learning objectives are rarely reflected in assessments, teacher standards or accountability frameworks. Teacher education and PD often ignore climate change, leaving educators to improvise. Infrastructure planning and climate adaptation for education are only patchily integrated into sector plans and budgets. In systems terms, climate appears in discourse and documents, but not in the core steering mechanisms of schooling. This is what we have called a low-climate-learning trap: a stable configuration in which climate education is always important "in principle" but rarely prioritized "in practice," because signals, incentives and capacities are misaligned.
- **Inequalities of exposure and opportunity**: There are also stark equity gaps. Children in climate-vulnerable regions (often low-income countries, rural areas, informal settlements) experience the most severe first-order disruptions but have the fewest resources for adaptation and the least access to rich climate education. Within countries, marginalized groups, including girls, Indigenous and minority communities, students with disabilities and those in under-resourced schools, are more exposed to hazards and more likely to attend schools that lack climate-ready infrastructure, curricula and support. Access to high-quality climate change education, especially programs that are interdisciplinary, justice-oriented and action-focused, is often concentrated in better-resourced schools and networks. Without deliberate strategy, climate change education risks reproducing or deepening inequalities: "green" opportunities for some, heightened vulnerability for others.
- **Evidence and monitoring gaps**: Although the evidence base is growing, there remain important knowledge gaps. Most studies linking education to environmental attitudes and behaviors are correlational. Rigorous causal and longitudinal studies, especially in low- and middle-income contexts, are few. Monitoring systems for climate-related learning are weak. Many countries lack indicators and data on climate in curricula, teacher preparation, student learning, school emissions or resilience. Evaluations of whole-school, network and cross-sector initiatives often focus on outputs (numbers trained, activities held) rather than outcomes for learning, behavior or climate resilience. These gaps constrain policy learning. Systems cannot improve what they cannot see; nor can they compare alternative strategies without better evidence.

9.3.3 Opportunities for Alignment and Transformation

Despite these challenges, the book has identified concrete opportunities.

1. Make climate a structuring theme of quality education. Climate and sustainability should not be treated as add-ons but woven into core definitions of educational quality and purpose. This is already implicit in global frameworks; it now needs to be explicit in national standards, curricula and sector plans.

2. Align key levers. Curriculum, assessment, teacher education, school leadership and infrastructure must send consistent signals about the importance of climate-related learning. Without such alignment, even excellent pilot programs will remain marginal.
3. Invest in educator and leader capacity. Teachers and school leaders are central agents of climate change education. They need sustained opportunities to build content knowledge, pedagogical skills, emotional resilience and leadership capacity, linked to their everyday work and supported by networks.
4. Leverage schools as community hubs. Schools already act as anchors for nutrition, health, childcare and crisis response. With intentional design, they can also become hubs for climate resilience and learning, hosting shelters, disseminating information, modelling sustainable practices, convening local planning.
5. Support and connect "niches". Informal networks, innovative schools, community programs and youth movements are already pioneering forms of climate education that are more integrated, justice-oriented and action-focused than system averages. Policy can protect, learn from and scale these niches, rather than ignoring or co-opting them.
6. Center justice, plurality and locality. Climate change education is most powerful when it takes seriously which communities are most affected, whose knowledge counts and whose futures are imagined. Integrating Indigenous and local knowledge, addressing gender and other structural inequalities, and connecting global science to local realities are not optional extras; they are conditions of relevance and legitimacy.
7. Build better feedback loops. Developing meaningful indicators, embedding evaluation in programs and sharing findings across sectors and borders can support adaptive improvement. This includes paying attention to unintended consequences, such as exacerbating anxiety or exclusion.

9.4 Implications for Different Actors

The book's analysis points toward differentiated, but interconnected, implications.

For governments and education authorities, the challenge is to treat climate change education as a core strategic priority: embedding it in sector plans and budgets; revising curricula, assessment and teacher policies accordingly; coordinating with environment, health, labor and planning ministries; and prioritizing support to the most climate-vulnerable regions and learners.

For schools and educators, it is to use available room for maneuver, however constrained, to integrate climate themes into teaching; to participate in and build professional networks; to work with students as partners in learning and action; and to use their institutional voice in local resilience and climate-policy conversations.

For multilateral organizations and donors, it is to continue refining frameworks and tools, but also to channel resources toward implementation, capacity-building and rigorous evaluation, especially in under-resourced settings, and to ensure that education is visible in climate finance and climate-policy processes.

For civil society and youth movements, it is to sustain pressure for ambitious, justice-centered climate-education policies; to develop and share models of practice; and to hold systems accountable for moving beyond rhetoric.

For researchers, it is to work in partnership with practitioners to design and study interventions that are theoretically grounded, context-sensitive and feasible to scale; to strengthen causal and longitudinal evidence; and to attend to ethical and emotional dimensions of climate learning.

9.5 Education's Role in a Climate-Changed Century

The book has been careful not to claim more for education than it can plausibly deliver. Education systems did not create climate change, and they cannot, by themselves, resolve it. The drivers of the crisis lie in energy systems, land use, economic structures, political choices and technological trajectories that extend far beyond schools.

At the same time, education is one of the few institutions that reaches most children and young people, is explicitly charged with imagining and shaping the future, can legitimize certain forms of knowledge, value and agency; and can connect individuals to wider publics and institutions.

In that sense, how education systems respond to climate change will influence which futures remain possible. If systems continue current trajectories, treating climate as a peripheral topic, allowing climate impacts to erode learning, leaving climate education to scattered enthusiasts, education will be part of the problem: reproducing patterns of denial, delay and unequal vulnerability.

If, instead, systems take seriously the insights presented in this book, about first- and second-order effects, systems and timescales, multiple narratives, promising practices, gaps and opportunities, education can become a meaningful part of the response: not a silver bullet, but a force multiplier for mitigation, adaptation and justice.

Doing so will demand the very capacities we hope to cultivate in learners: the ability to see systems, hold tensions, act collectively, and sustain hope without illusion. It will require educators, policymakers, researchers, families and young people to recognize themselves as participants in intertwined education and climate systems, and to exercise agency, within their spheres of influence, to shift those systems toward more sustainable and humane paths.

The work is daunting and necessarily unfinished. But it is also, fundamentally, educational work: learning, together, how to live and learn well on a rapidly changing planet.

Open Access This chapter is licensed under the terms of the Creative Commons Attribution-NonCommercial-NoDerivatives 4.0 International License (http://creativecommons.org/licenses/by-nc-nd/4.0/), which permits any noncommercial use, sharing, distribution and reproduction in any medium or format, as long as you give appropriate credit to the original author(s) and the source, provide a link to the Creative Commons license and indicate if you modified the licensed material. You do not have permission under this license to share adapted material derived from this chapter or parts of it.

The images or other third party material in this chapter are included in the chapter's Creative Commons license, unless indicated otherwise in a credit line to the material. If material is not included in the chapter's Creative Commons license and your intended use is not permitted by statutory regulation or exceeds the permitted use, you will need to obtain permission directly from the copyright holder.

Index

© The Editor(s) (if applicable) and The Author(s) 2026

F. Reimers and M. Wang-Aghania, *Educating for a Climate Changed Future*,
https://doi.org/10.1007/978-3-032-23256-4

C

G

H

N

O

P

T

U

Y

Z

FSC
www.fsc.org
MIX
Papier aus verantwortungsvollen Quellen
Paper from responsible sources
FSC® C105338

If you have any concerns about our products,
you can contact us on
ProductSafety@springernature.com

In case Publisher is established outside the EU,
the EU authorized representative is:
Springer Nature Customer Service Center GmbH
Europaplatz 3, 69115 Heidelberg, Germany

Printed by Libri Plureos GmbH
in Hamburg, Germany